Engineering

A Beginner's Guide

ONEWORLD BEGINNER'S GUIDES combine an original, inventive, and engaging approach with expert analysis on subjects ranging from art and history to religion and politics, and everything in between. Innovative and affordable, books in the series are perfect for anyone curious about the way the world works and the big ideas of our time.

anarchism
artificial intelligence
the beat generation
biodiversity
bioterror & biowarfare
the brain
the buddha
censorship
christianity
civil liberties
classical music
cloning
cold war
crimes against humanity
criminal psychology
critical thinking
daoism
democracy
dyslexia
energy
engineering
evolution
evolutionary psychology
existentialism
fair trade
feminism
forensic science
french revolution
history of science
humanism
islamic philosophy
journalism
lacan
life in the universe
machiavelli
mafia & organized crime
marx
medieval philosophy
middle east
NATO
oil
the palestine–israeli conflict
philosophy of mind
philosophy of religion
philosophy of science
postmodernism
psychology
quantum physics
the qur'an
racism
the small arms trade
sufism

Engineering
A Beginner's Guide

Natasha McCarthy

ONEWORLD
OXFORD

A Oneworld Paperback Original

Published by Oneworld Publications 2009

A CIP record for this title is available from the British Library

ISBN 978–1–85168–662–9

Typeset by Jayvee, Trivandrum, India
Cover design by www.fatfacedesign.com
Printed and bound in Great Britain by Bell & Bain Ltd., Glasgow

Oneworld Publications
185 Banbury Road
Oxford OX2 7AR
England
www.oneworld-publications.com

Contents

Preface

What is an engineer? What comes to mind when imagining an engineer? Someone who gets the train going when it's stuck in the station? Someone who comes in and fixes your washing machine when it refuses to drain away the murky grey water?

Maybe it is more appropriate to think of someone who designs and oversees the construction of the highly complex piece of machinery that is sitting in the station waiting to set off. Or even someone who designs and manages the transport infrastructure of which the station or stretch of railway at which you're stuck is but a fragment. It might be the person who designs the complex construction of pumps and logic-controlled programming that washes your clothes, or who manages the international enterprise of designing and constructing white goods for a particular company.

Perhaps you might even imagine the telecommunications or software expert that makes your train travel unnecessary through the provision of teleconferences or virtual discussion boards. Or the chemical engineer who has eradicated the chore of clothes washing, by … . Actually, we are still holding out for the latter, but here's hoping.

Engineering encompasses an extremely broad range of activities on a whole spectrum of levels. Engineers are responsible for the design, production, delivery and maintenance of everyday objects such as cars, PCs, telephones and vacuum cleaners. They are also responsible for the design and production of not-so-everyday objects such as space shuttles and kidney dialysis machines. And they are responsible for things that are not objects at all in any straightforward sense – from road and rail systems, to the networks of pipes and wires that deliver water and electricity, to the cellular networks that support mobile phones, to the IT systems that process millions of financial transactions each day. Engineers work at many levels, from designing the smallest component of a device, to the management of whole design projects, to overseeing construction sites and production lines. Not all of those people that might be called engineers

ENGINEERING AT THE HAIRDRESSERS

It is easy to get a sense of how pervasive engineering is by looking at your setting, at any given time and considering how many of the things around you are the products of engineering. For example, this morning I was at the hairdressers. There are probably few places one can think of that are further from the world of engineering (unless perhaps certain hairdressers take to calling themselves hair engineers) yet the modern hairdressers is completely dependent on the products of engineering. At the basic level, a modern hair salon is dependent on the supply of electricity – impossible without electrical engineering – and of water – impossible without civil engineering, and boilers are needed to heat the water – designed by electrical or gas engineers. Hairdryers are the product of electrical engineering – equipped with electrically driven fans and heating elements. Modern hairdressers make use of styling tools that use ceramic coatings – which conduct heat rapidly and have a smooth texture; and which are a product of materials engineering. Paying for the haircut involves – like paying for many services in the UK and other countries now – use of chip and pin technology, in which the card reader communicates with a chip in your credit card to assess whether the correct pin has been entered and which then wirelessly transmits information about the transaction to your bank; a wonder of modern communications engineering and encryption technologies. Of course, barbers and hairdressers existed long before modern engineering did, but the modern hairdresser is completely dependent on the achievements of engineering.

are agreed by the profession to be engineers, but as a label it probably beats 'scientist' for its breadth and diversity.

As a result of this breadth, the task of writing a book on engineering might seem to be near impossible. In particular, an engineer working in one area of engineering, appreciating all the detail and complexity that there is to cover in just the area of, say, wastewater management, will probably find it unimaginable that in one short book one could write an introduction to the *whole* of engineering. Perhaps it is because I am not an engineer immersed in a particular area or project that I am not put off the whole idea! This book is written from the outside of engineering looking in – looking at engineering in its broadest and widest sense; making sweeping generalisations perhaps, but with the aim of capturing the huge, broad, all-encompassing effect on the world that the practice of engineering has had. Hopefully this approach will convey the sense of awe of someone who has not been involved in the nitty gritty, but who has worked with many people who have.

My intention is that the non-engineer will get from this beginner's guide a sense of the role of engineering in shaping lifestyle, culture, society, knowledge – as great a role perhaps as the more often lauded sciences and arts. I hope that the engineer reading this will get a novel and inspiring perspective on the whole of engineering, a context in which to set the detailed work and local problems that they deal with on a daily basis. Engineers care deeply about their profession and I hope that this book conveys some of the belief that engineers have in what they do and its importance. It is also the purpose of this book to show that engineers and engineering have a vital role to play in figuring out ways to support our quality of life long into the future. The impacts of climate change, for example, will affect everyone across the globe and strategies to slow its progress and mitigate its effects are needed desperately, be that in terms of cleaner ways to produce energy or ways of making essential activities – heating and cooling buildings, travelling, computing and so on – far more energy-efficient. While politicians across the world agree on targets for cutting carbon emissions and forge agreements to tackle climate change, it is engineers who will have an essential role in turning promises and good intentions into reality.

The following chapters will take a tour through the many achievements that are the product of engineering, the impression that engineering has made on the world we inhabit, and the impact (for good or ill) that the engineer makes today and in the future. This is not an introduction to engineering in the sense of being a potted account of basic engineering practice. There will be no beginner's guide to constructing a steel-framed building or designing surgical robots, because that is simply not possible (and any attempt would invoke a whole heap of health and safety liabilities)! Rather it will seek to give an overview of what, if anything, is characteristic to engineering. It will look at the kinds of problems that an engineer faces, the ways that engineering deals with those challenges and the contribution of engineering to human knowledge through its endeavours to improve our way of life. It will be a celebration of the often overlooked achievements of engineering, a critique of the harm engineering practice may have caused, and a challenge for the engineers of the future. Engineers will have a major role in addressing some of the biggest challenges facing the world.

The book is written from a social and philosophical perspective and will raise as many questions about the role and responsibility of an engineer as it answers. I hope that this will make it accessible to the non-engineer (and the non-mathematician and non-physicist) and will show the technical and non-technical reader alike what a great deal there is to learn from engineering.

Acknowledgements

I would like to thank the Fellows and staff at The Royal Academy of Engineering who have provided inspiration for this book. Particular thanks go to Tony Eades, Brian Doble, Keith Davis and the policy team for their encouragement. I am extremely grateful to the following people for their valuable comments on various sections of the book: Igor Aleksander, David Andrews, Brian Davies, Michael Davis, Christopher Elliott, Allan Fox, Nigel Gilbert, David Goldberg, Dame Wendy Hall, Ian Howard, Nick Jenkins, Christopher Kent, Sir Duncan Michael, John Monk, Raffaella Ocone, Richard Ploszek, John Roberts, Christopher Snowden, Martyn Thomas, John Turnbull, Alan Walker and John Yates.

This book is dedicated to Stuart and his affinity for science and technology.

Illustrations

1

The evolution of the engineer

Some activities have a clear relationship to their ancient predecessors. Mathematics is one example. However much it has developed since then, ancient activities that involved dealing with quantities and ratios through numbers or other abstract means count as mathematics today. The same goes for philosophy. Anyone who spent a significant part of their day considering the meaning of life and how to live it – either strolling through the marketplace like Socrates or sitting alone in a barrel – counts as a philosopher. However naïve and uninteresting the resulting philosophy, it is still a contender as proto-philosophy.

Engineering is not so clear cut. It is obvious that activities resembling the aims and purposes of engineering have been around since Homo erectus first started to build shelters. We might consider the first uses of flints to cut or create fire as examples of early engineering, but is it right to count stone-age tool makers as engineers? Since we now distinguish between engineering and manual arts and trades this does not seem wholly accurate – after all, not everyone who masters the use of tools is an engineer; dabbling in home decoration however fancy your power tools is not sufficient. Whilst we call it mathematics when a child subtracts numbers of apples from numbers of pears, however different it is to the work of a professor spinning out proofs in his office, we do not call amateur construction work engineering.

So when should we say that real engineering emerged? There are a number of ways to tell this story, and all depend on arguable decisions about what constitutes an engineer.

How old is engineering?

One way to map out the history of engineering is to trace it back through all of those activities that significantly resemble what we would now count as engineering. Activities can be similar to engineering for a number of reasons. For example, there is nothing in existence now like the Egyptian pyramids and no living engineers work on similar projects. However, the building of those edifices involved the application of mathematical relationships to create a physical structure and required complex, large-scale organisation of manpower and resources. Both of these are characteristic of contemporary engineering. However, the Egyptian pyramids seem to differ somewhat from modern engineering in that engineering projects are focused on meeting some useful purpose that can serve humanity and improve quality of life. The pyramids, impressive though they are, have an other-worldly purpose that is rarely present in modern engineering endeavours.

In ancient Greece there were many activities that seem to be precursors to what we might now call 'engineering science' – the study of the mathematical and physical principles underlying the structures and machines that we build and use. For example, Archimedes developed an analysis of how levers function, and he was so confident in his work that he believed he could apply it to position a lever that would move the earth itself. However, the ancient Greeks were not so keen on getting their hands dirty, preferring rather to sit and contemplate the nature of fulcra and pulleys. Therefore, such work is perhaps better characterised as applied or applicable mathematics than as engineering. The

Romans did rather better, focussing their attention on more practical matters such as building viaducts and sewage systems. Though perhaps not done with as much application of mathematical principles as modern engineering, this work surely deserves to be called engineering of some kind, even if rather rudimentary.

Activities of the medieval period, or the unfairly labelled 'dark ages', have a close link with modern engineering. The kinds of structures built in medieval Europe, vast vaulted cathedrals which ingeniously incorporate structural support into their decoration through the use of devices such as flying buttresses, can be compared to ambitious structural engineering projects. Although the craftsmen behind their construction were described as master builders, and might be seen as similar to modern architects, their work certainly encompassed engineering tasks. After all, nowadays one would not entrust such grand and ambitious structures to architects and construction workers alone without the input of skilled structural engineers. At the same time, the Islamic Golden Age brought with it great leaps in mechanical invention. A central figure of this period was Al-Jazari, an inventor and craftsman who published his life's work in his *Book of Ingenious Mechanical Devices*. Among those devices were mechanisms that were crucial to mechanical engineering for many centuries, such as water-raising pumps, and sophisticated curiosities such as hand-washing automata.

In all of these periods there are certainly activities that foreshadow the work of contemporary engineers. However, it is difficult to map out the story of engineering in terms of what people do, as many activities can resemble engineering in many ways. Physicists and chemists do work that is similar to engineering in some ways, but we would not call them engineers. It is interesting, therefore, to trace the engineering discipline back through the etymology of the word 'engineer', to see when the discipline first got its name.

The word 'engineer' is generally claimed to be rooted in the Latin term *ingenium* or *ingeniatorum*, meaning, respectively, *ingenuity* or *one who possesses or exercises ingenuity*. This is, of course, the same root from which the English word 'ingenious' springs, meaning inventive and novel, implying that central to the concept of engineering is that of inventiveness and creativity. In the medieval period, those craftsmen working in the military on catapults, or other devices of war, were known as 'ingeniators'. A version of the title persisted in the renaissance period, with Leonardo Da Vinci proudly bearing the title '*ingeniarius ducalis*' at one point in his career. Does this title indicate that he was an engineer in the modern sense? His work, as described in the box below, certainly encompasses engineering activities, and engineers are often keen to claim him as one of their own. However, the reasons for being called 'ingeniarius' are slightly different from the criteria we use for calling someone an engineer today.

The word 'ingenium' is also the root of the English word 'engine'. We might assume that engineers are so-called because they work with engines, but of course the medieval 'ingeniator' did not work with anything like modern engineering machines such as steam or combustion engines. In the medieval and early modern times, 'ingeniators' were so-called because they worked in the military, with 'engines' of war. This category covered all manner of early war technology, such as catapults and cannons, and ingeniators would devise, build and maintain these. Indeed, Leonardo Da Vinci's engineering work focuses heavily on designs of guns and other bombardments – the renaissance machinery essential to win out in a siege. So the title 'ingeniator' primarily relates to work in the military rather than general engineering.

The English word 'engineer' made its direct entry into the language through the French word 'ingénieur'. Here, again, the original uses of the word apply to activities within warfare. The ingénieurs in the French military were still those members of the army who dealt with and maintained engines of war, but in the French army they had a rather lofty status compared to their earlier counterparts. They were, in 1676, formed into the *corps du génie*, a group of soldiers who received special training in military construction and who began to hone the art of making engines of war and creating the infrastructure – roads, fortification and such – that the military required. Again, the word 'ingénieur' denoted a certain member of the military with specific skills and abilities.

The term 'engineer' now refers to a much wider class of people than those who work in support of the military. Therefore there are weaknesses in telling the story based on the name 'engineer', as it does not tell us when the wide-ranging discipline that we now recognise as engineering sprang up. However, it is undeniable that military activities had a significant role in shaping the modern engineers, as is seen by looking at the history of engineering education and training.

LEONARDO DA VINCI

Leonardo Da Vinci was a painter – a painter of such skill that his name and works live on as probably the most famous in the Western world. However, he was a man of the renaissance, dabbling in and even excelling at a wide range of disciplines; in Leonardo's case this included engineering. Leonardo identified himself as an engineer. There is evidence of a letter of introduction by Leonardo in which he sets out in detail his engineering skills. Military engineering skills feature heavily – producing designs for catapults and cannons, and developing methods for making wrought metal to manufacture cannons in novel ways for added strength. However, also included are bridge design and town planning; hydraulics, including studies on taming the course of rivers; and building automata for entertainment at grand events. It

LEONARDO DA VINCI (*cont.*)

is well known that Leonardo designed flying machines and a helicopter and there is evidence that he even trialled these – but to no success.

Leonardo's notebooks containing his thoughts and ideas on engineering are incomplete and it seems that many of his thoughts were too. Hence, he has not made the impact on the history of engineering that, say, Galileo did in his work on engineering science. Some have even argued that he was a fairly ordinary engineer and that he created nothing that did not have precedent in the work of his peers or predecessors.[1] Perhaps it was his average performance in this area compared to his genius in the area of painting which means that this career has completely eclipsed his engineering efforts. Or perhaps it is because modern technologies – in particular aircraft, helicopters and submarines – have completely surpassed any work he did on these areas, whilst no advances in art (if there are such) could ever show the Mona Lisa to be a primitive or poor painting. Whatever the reason, it is interesting to consider whether Leonardo would be surprised, even disappointed that this was his primary legacy. It is certainly noteworthy that the renaissance differs from the current age in terms of the status and perception of engineers. Whereas now, to be a successful artist is considered noble and noteworthy and to be an engineer is thought relatively modest, in Leonardo's time to make and to imagine things of a technical nature was as lofty as creating fine art.

The French military of the later seventeenth century was the birthplace of engineering as a distinct profession with a specialised mode of training and education. The *corps du génie* was a body of soldiers in which apprentices, taught in training camps, were instructed in the specialised arts of military technologies. This was established in 1676 and was followed 30 years later by the *corps des ponts et chaussées* (bridges and roads). This latter body focused on establishing the infrastructure needed by both the military and civilians. A school was established in 1747 to train students specifically in the arts of civilian engineering, the *École des Ponts et Chaussées*, which still takes students now (known as the ENPC). However, the first school to teach a curriculum that bears a close resemblance to the engineering education we see now was the *École Polytechnique*, set up in 1794 as the *École Centrale des Travaux Publics*.

The French engineering education was way ahead of the rest of the world; it was the only country that was teaching engineering in a principled way, away from the everyday work of the job. The US was quick to follow, however, and engineering schools were developing out of military training schools, beginning with the establishment of West Point (now United States Military Academy at West Point). Some years after its establishment it took on the curriculum of the French military engineering schools and by the 1830s numerous successful engineering schools were developing in the US that taught some version of the same curriculum.

The United Kingdom was slow to follow, however. Throughout the Eighteenth century the UK had no specialised training for engineers either in the military or the civilian worlds. Engineers learned through apprenticeships – being trained by observing and assisting an accomplished engineer. Only for some engineers did this on-the-job training follow a university education. Well-known engineers such as Thomas Telford and James Watt started their careers as apprentices to craftsmen, learning practical skills rather than theory. However, despite their manual training, their role was that of an engineer. They were involved in design, in planning works for a client and contracting others to actually carry out the labour, rather than carrying out the manual work themselves. John Smeaton (1724–1792) was one of the first engineers to work in this way in the UK, and it was he who first referred to himself as a 'civil engineer', ascribing the title to himself on a report he prepared in 1768.

Once we get to Smeaton, we get to a point in history where we can uncontroversially identify an individual as a modern engineer, working outside of the military. However, the stories above show that there are a number of different paths to this point in history, and a number of stopping points at which engineering might be argued to have emerged. They demonstrate that retelling the history of engineering is a difficult matter. Engineering was not born fully fledged at some distinct point in history but evolved over an extended period of time. At different points during that period, in different parts of the world, activities close to modern engineering can be spotted, and individuals identified who come close to modern engineers, but it took some time for the distinct discipline to emerge.

The emergence of a profession

The long process of engineering's evolution was followed by a more rapid phase of self-organisation in engineering. As engineering activity became more intensive and the ambitions of engineers greater, engineers felt the need to establish themselves, and to be recognised in the public eye, as a profession. A profession brings with it a body of agreed knowledge shared by members. It usually involves specific training or qualifications, and it requires its members to commit themselves to shared standards of work and agreed ethical principles. While the emergence of an engineering education in France and the US was therefore central to engineering becoming a profession, another major milestone was the establishment of a professional engineering body.

The lack of a formal education system for engineers in the UK left some young engineers in need of a source of information and education that could support them in their development as engineers. It was this need that led Henry Palmer to organise a meeting on Friday 2 January 1818 in Kendal's Coffee House, Fleet Street, London. Palmer's opening speech gave the following view of engineering:

> The Engineer is a mediator between the philosopher and the working mechanic and like an interpreter between two foreigners, must understand the language of both, hence the absolute necessity of possessing both practical and theoretical knowledge.

This illustrates the unique nature of the engineering profession that was emerging. It was more than just the manual work of the craftsman or labourer, but unlike pure science it was directly related to the practical tasks of the artisan. This new kingdom needed a constitution and the Institution of Civil Engineers (ICE) was the first organisation to fill this role. It was initiated by a group of young men whose ambition was to provide an opportunity for engineers to receive instruction and guidance, and who were willing to share their growing experience and expertise in order to achieve this. The group appointed Thomas Telford as its first president. Telford was an engineer of significant reputation and influence, who had spent his career building roads, canals and bridges. His most famous works include the Caledonian Canal, which created a navigable route across the entire width of Scotland from Inverness to Fort William, and the Menai suspension bridge in Wales. Telford donated his library to the Institution and lent it his reputation and wealth of experience. With Telford at its head the ICE became established as a body of great significance. The ICE was a model that was soon to be copied elsewhere. In the US, the first professional body for engineers was the Boston Society of Civil Engineers established in 1848, with the American Society of Civil Engineers following in 1852. Similar professional organisations are now established world wide with some countries controlling the title 'engineer' so that no one can practice as an engineer without being registered with the appropriate professional body.

The founding of the ICE not only established momentum for similar institutions to mobilise across the world but it set in place an explosion of engineering societies representing ever new and more specialised areas of engineering. The first branching occurred when the Institution of Mechanical Engineers was set up in 1846. Part of the reason for taking their own path was the feeling that mechanical engineers were quite different to civil engineers. While the consulting civil engineer might spend time drawing plans and visiting sites to ascertain that all is going to plan, the mechanical engineer is more likely to be getting dirty hands in the cabin of a locomotive. Hence the mechanical engineers created their own society serving their needs more precisely. This was a move that has often been repeated – and following the initial distinction between civil and military engineering, there have come many distinctions that can be made at more or less detailed levels.

Throughout the world, engineering has followed this model, with specialised engineering societies being established to represent the interests of their members and to set standards of practice and ethics. The emergence of engineering as a profession and its organisation into professional bodies puts engineering on a par with other professions such as medicine and law. It reflects the fact that engineering requires specialist knowledge and skill for the engineer to work successfully, and that engineering has a central role in society, serving and bearing responsibility to the wider public.

ISAMBARD KINGDOM BRUNEL

Isambard Kingdom Brunel is one of the most famous names in engineering. Brunel was born on 9 April 1806, the son of Marc Brunel and Sophia Kingdom. Marc was an engineer who had made important developments in the process of mass manufacturing and recruited his son on one of his greatest works, the tunnel under the River Thames in London. I. K. Brunel was a precocious recruit and his successes saw him join the scientific elite in Britain's Royal Society at the age of 24.

Brunel's greatest successes lie in his railways and bridges. Brunel had a hand in the construction of more than 1200 lines of railway in Britain. The line that is most closely associated with him is the Great Western Railway (GWR) from London to Bristol. Paddington Station was designed in outline by him, with the detailed design and construction handled by the same company that built the Crystal Palace for the Great Exhibition in 1851. Brunel's railway lines were always created with an eye on aesthetics, and the bridges and tunnels of Brunel's railways were carefully planned by Brunel himself. Two bridges that stand out as his legacies are the Royal Albert Bridge in Saltash, Cornwall and the Clifton Suspension Bridge across the Avon Gorge in Bristol. The first stone of this bridge was laid in 1831, however work was halted, and only taken up again after his death as a memorial to him (though the final bridge was of a slightly altered design). Brunel developed mathematical analyses for application in the design of his bridges and had, unlike his compatriots of the time, a grounding in engineering science from his education at schools intended to prepare students for France's Ecole Polytechnique.

Brunel did not always meet with such success. His steamship The Great Eastern was designed to be larger than any predecessor ship in order to take passengers in luxury on a non-stop journey from Bristol to the Far East – carrying all its fuel with it. The construction of the Great Eastern was a troubled affair from construction to launch. It did not serve for any great period of time as a passenger ship, as demand did not exist, but ended up being used to lay undersea cables for telegraph wires.

The great Eastern was a financial disaster, losing money for the

ISAMBARD KINGDOM BRUNEL (*cont.*)

companies that financed it, that built it and even broke it up for scrap. Was there anything successful about the Great Eastern? It failed to meet its intended purposes and did little to improve human welfare. However, as a piece of engineering it was a huge development, and its failure lay mostly in the fact that it was ahead of its time – there simply wasn't a need yet for such a huge ship. But should an engineer work on something for which there is no demand or need?

Brunel was a proud man and was often mistaken about the scope of his abilities or the wisdom of his grand plans. Two examples illustrate this well. The first was the 'atmospheric railway' he built in Devon – a railway on which trains are pulled along the tracks by pistons that draw up inside hollow tubes running the length of the railway, 'sucking' the train along. Unfortunately, the system did not work in practice, with air constantly leaking into the tubes. Brunel also erred when he commissioned locomotive engines to his own design to run on the GWR. These were built but were hopelessly substandard and had to be replaced. Brunel simply overestimated his own ability.

It is arguable that in some areas of his work, Brunel did not quite get the success he deserved. Brunel's GWR was built using 'broad gauge' track – 7 feet and one quarter of an inch wide. However, most of the railway in the UK at that time used track of 4ft 8.5 inches. This was used by the Stephensons, Brunel's friendly adversaries, as it was the width of track used for horse-drawn railways and would fit with pre-existing tracks in mines. Brunel however selected the wide gauge on the basis that it gave a smoother ride. Although this would mean problems of compatibility with the narrow gauge Brunel was insistent on his railway being the finest in Britain. However, as the railway system grew in the UK one size had to win out. Although it was demonstrated that the broad gauge track allowed locomotives to pull greater weights at faster speeds with less discomfort, by the time a decision was made there was six times as much track in narrow gauge than broad gauge. For consistency and simplicity's sake, what was possibly the lesser system from an engineering point of view won out.

What makes someone an engineer?

The fragmentation of engineering societies into ever more specialised groups shows that the evolution of engineering did not stop with its self-organisation. The engineering profession is like a life form that keeps evolving into distinct species as new niches appear. Developments in technology and even changes in society often lead to the creation of a new, distinct branch of engineering. Nuclear engineering developed as a profession in the mid-twentieth century when nuclear physics became sufficiently mature to allow the controlled use of nuclear reaction to generate electricity. Biomedical engineering has developed as the potential to apply technology in medicine has grown. It is a natural process that, as knowledge grows and technologies improve, specialists are needed that are expert in the detail of these technologies. Education and training is required in these specialisms, and thus organisations are created as a home for these individuals and a resource for their training.

So what is it that unites the engineering profession? What makes someone an engineer? What must they do, or know, or have learnt to legitimately count as an engineer? There is no shortage of delightful soundbites that attempt to encapsulate the essence of engineering. Here are two. The original charter of the Institution of Civil Engineers drawn up in 1828 defined engineering as 'the art of directing the great sources of power in nature for the use and convenience of man ...' Less concisely, but still impressively, Herbert Hoover describes the experience of being an engineer thus: 'It is a great profession. There is the fascination of watching a figment of the imagination emerge through the aid of science to a plan on paper. Then it moves to realisation in stone or metal or energy. Then it brings jobs and homes to men. Then it elevates the standards of living and adds to the comforts of life. That is the engineer's high privilege.'[2]

There are a number of characteristics of engineering that are clear from these sketches. It is worth drawing each out and considering how they are reflected in engineering practice and what they demand of the engineer in terms of knowledge and skills.

It is striking that engineering is often explicitly characterised as existing to serve humankind. It might be surprising to the non-engineer that this is so central to the aims of engineering institutions, but engineering is fundamentally concerned with creating technologies that will be of use to people, will improve their welfare and raise their quality of life. Engineers might not stereotypically be seen as driven by compassion for others, an ambition reserved perhaps for caring professions such as medicine. However, few engineering projects have any meaning apart from making peoples' lives easier. From modern conveniences such as mobile phones, to sewage systems that have always been essential for people to live healthily in communities, engineers create systems that make human existence easier, healthier and more efficient. Of course, this is not all altruistic. It is identifying what makes peoples' lives easier that allows engineers to turn their work into viable business.

The fact that engineering is so focused on human needs means that engineers need to have an understanding of the nature of human needs and desires, and how to create artifacts that successfully meet those needs and which are easily used. This is quite some challenge, especially given that few engineers study social sciences such as psychology and sociology that might give them a deeper insight into human wants and needs. Whether engineers need this facet to their education will be explored later.

A focus on human needs is not the territory only of the engineer. What makes an activity engineering is that it meets human needs by changing the world that humans inhabit.

Engineering might involve extracting natural materials from the earth and processing them so that they can be used to fulfil human needs – for example, oil and gas for fuel, stone, wood and metal for producing structures and artifacts. Engineering might otherwise involve exploiting the laws of nature to create artifacts and systems that allow us to create power and harness energy – from the steam engine to aircraft. Engineering might also involve harnessing both natural materials and natural laws to produce new materials.

Therefore engineering is inherently creative – it takes what nature provides in order to make something new, or to make things behave in novel ways. Engineers should never be satisfied with the way things are; the successful engineer will always want to intervene, to change things so that they work better.

The engineer's creativity is of course naturally constrained by the resources that the world provides, the properties of those resources and the limits of the laws of nature. Thus, insight into the nature of materials and physical law are essential to engineering's sophistication and growth. Engineering builds on, and demands an understanding of, fundamental sciences such as physics and chemistry. It is the appreciation and application of scientific theory that distinguishes engineering from crafts, or more practical and less sophisticated manual pursuits. For centuries humans have used the resources around them for food, clothing and shelter, but it is the methodological, self-conscious and scientifically-based adoption of nature's materials and laws that distinguishes engineering from all manner of practical pursuits from the cave to the cottage industry.

All of these elements are crucial to engineering and form the core of what counts as engineering. But to fully characterise the concept of engineering it is helpful to look not only at what is central to it, but at what happens at its borders. Engineering sits alongside technology, science and mathematics – but what makes it different to these other pursuits?

It *is* rocket science actually ... engineering 'versus' science

Engineering and the sciences are obviously closely aligned. Engineering is frequently depicted as being dependent on science, this dependency being implicit in the description of engineering as 'applied science'. Categorising engineering as applied science could suggest that its role is secondary, finding a use for the discoveries of science rather than making discoveries of its own. It brings with it the suggestion that engineering is somehow more prosaic and less creative, involving more donkey work than brain power.

The view of engineering as applied science and the relationship that it implies sometimes makes engineers feel hard done-by. Science seems to hog the limelight and even gets the credit for the ground-breaking work that is actually the result of engineering. For example, engineers often balk at the expression 'it's not rocket science', pointing out that 'it' is indeed not rocket science, it's rocket *engineering*: engineers, not scientists, constructed the space shuttles that allow space exploration and design and build the satellites that populate our planet's orbit. Of course, rocket engineering (the unofficial name for a branch of what is officially aerospace engineering) does depend to a great extent on physics. But does this mean that engineers are simply applying the work that physicists do? This suggests that the engineers simply take scientific theory and follow it, recipe-like, to produce a working artifact. The actual design process in engineering is far from simple, involving both appreciation of the scientific theory and its limits and an understanding of what applications are needed and are likely to be successful. Finding applications for scientific theory and adapting knowledge for useful outcomes is itself highly creative.

Engineering is in fact a quite different entity to science, exploring the world for different reasons and by different means.

ROCKET MEN: HOW ASTRONAUTS NEED ENGINEERS

Knowing how to get a person into space and actually achieving that aim are light-years apart. Although it is impossible to consider devising a vehicle for space travel without an underpinning body of physics to guide the design, the achievement would be impossible without engineering. The basic science behind rocket engineering is not cutting edge, but in fact is based on Newtonian physics, dating to the seventeenth century, which provides an understanding of the effect of gravity on a body in motion.

One of the central challenges of space travel is that of overcoming the power of earth's gravity, the massive pull that attempts to keep objects close to its surface. Once that is defeated, a space craft can travel more easily, eventually moving freely in an orbit of the earth. A space vehicle requires massive thrust to overcome the force of gravity, and this comes in the form of booster rockets that provide the propulsion that will carry a satellite or space vehicle into orbit or beyond. The technology for space flight was developed from rocket technology used to build missiles in the Second World War, where the research and development was focused on giving missiles enough propulsion to travel to such a height that their guided fall would allow them to travel significant distances (and then, of course, to cause death and destruction where they landed). A huge part of this challenge was the theoretically simple but practically significant feat of building a rocket, powered with the highly volatile fuel needed to provide the required thrust, in a way that would not destroy the rocket and its cargo in the process – a target that was missed on many attempts.

Space exploration is therefore simply impossible without engineering. Without denying for a moment the sheer guts it must have taken for early space pilots to take those first extremely risky journeys into the unknown, it was the dreams and efforts of a team of highly skilled engineers that turned vision into reality. Chief amongst these were Wernher von Braun (1912 –1977), a

ROCKET MEN: HOW ASTRONAUTS NEED ENGINEERS (*cont.*)

German rocket engineer who took his skills and secrets to the US after the Second World War, and Sergei Korolev (1907–1966), a one time prisoner in the Gulags who worked on the Soviet rocket programme which culminated in the rockets that launched the first satellite into orbit and the first man into space.

Science aims at constructing and testing theories by observing the world and discovering new phenomena, thus better to understand the natural world; engineering is concerned with constructing artifacts on the basis of a cycle of design and testing, thus better to adapt and inhabit the natural world. This difference can be framed in a number of ways. Sunny Auyang puts it thus: 'Natural Scientists discover what was not known. Engineers create what did not exist.'[3] Other ways of drawing a line between the two identify the subject of science as being *how things are* whilst the focus of engineering is on *how things could be.*

A prevailing view of science is that it aims to describe the world, as it is, independent of the scientist. Science aims to spell out the secrets of nature only in order to tell a good story – one which, depending on your philosophical view, might be literally true, or a good interpretation of the observable facts. But the engineer wants to get at these secrets in order to use them, to play nature at its own game and to reshape the world for human use. The engineer wants to get to know the world in order to *change* it. If scientists changed the world in order to get to know it, they might be seen as failing in their attempt to depict the world as it *really* is.

To be more specific in drawing the line between science and engineering, we can say that the crucial activity that divides the scientist from the engineer is that engineers engage in *design.*

The following quote from Henry Petroksi demonstrates a common view: 'The idea of design and development is what *most* distinguishes engineering from science, which concerns itself principally with understanding the world as it is.'[4] Design is about systematically and methodically devising and creating something new. It is through the activity of design that engineers engage with the world that could be, rather than the world that is. Engineers seek to expand on or to improve what already exists; but they are constrained by the way that the world *could* be, that is, they must work within the parameters of physical and mathematical law. Thus their exploration of the world of the possible is different to that of the artist or storyteller. And unlike the storyteller, and even possibly the scientist, the engineer's work focuses on not just any way that the world could be, but on how the world could be *improved*. The engineer's aim is to create artifacts that can have a valuable role in people's lives.

It is important not to put too much weight on distinctions between science and engineering, however. The two are not the different sides of a coin but the ends of a spectrum, and much lies between the two extremes. The area of engineering science is rather more like physics than design. Engineering science is concerned with developing an analytical description of the human-made world – or at least aspects of it. It studies the behaviour of artifacts from bridges to aircraft, and the ways that they relate to the natural environment and are affected by natural laws. Its aim is to understand those parts of the world that have been directly affected by humankind, and its purpose is to provide understanding that will ultimately allow the better design and maintenance of human-made objects. Many areas of study lie on the borderline between natural science, engineering science and practical engineering. One such area is electronics, where an understanding of the physical phenomena, the building of electronic systems and their use in experimentation are closely intertwined. Researchers in electronics will find it a challenge to identify themselves as either engineers or physicists, as it is in such areas that the blurred boundary becomes so attenuated as to be almost non-existent.

Engineering and other arts

How do we distinguish engineering from technology? The two are often mentioned in one breath, and engineering is often implied in discussions of technology. Technology is a broad and inclusive concept. It can mean a set of objects, artifacts made by human design. Or it might refer to the process by which the artifacts came about and the knowledge needed to engage in that process. Writers on technology, its history and place in our lives frequently understand technology as encompassing the sum of all of human making and devising, dating back through cottage industry to crude inventions made in the cave. Histories of technology often start with early human production of tools as earliest examples of technology. Thus the kinds of activities ruled out at the start of this book as proto-engineering are classed as legitimate examples of technology.

Under this definition engineering has its place as a subdivision of technology, but would count equally as a form of technology alongside baking, tailoring and pottery. Engineering would stand apart, however, as that particularly sophisticated body of knowledge that produces particularly useful, ingenious and adaptable artifacts. Our interest in this book is in this more sophisticated end of the spectrum of human devising, and how this has changed human life. Technology broadly understood is fundamental to human existence, but engineering is relatively new, and has evidently brought changes to human experience.

In everyday language, the word 'technology' is used to talk about 'high technology', the state of the art in electronics. If one looks at the technology sections of a newspaper or website, the predominant theme would be the latest developments in electronic gadgets, those items not yet in the mainstream but objects of wonder and desire being picked up and tested out by a small band of devotees. In this sense, technology has a quite close relationship to engineering, but the two ideas are still quite distinct. Whilst high technology and the latest developments in any field often have a great deal to do with engineering, they tend only to overlap. Engineering has a part to play in developing high-end gadgets. But technology in this sense is also wrapped up in design, in marketing, in business and in the expertise of each of the specialist worlds in which new technologies are developed. What engineering brings to these areas is the application of the latest scientific developments to provide breakthroughs in each area. It is this focus specifically on using scientific knowledge to create breakthroughs that really sets engineering apart from other areas of inventing and making.

JONATHAN IVE AND APPLE

There is a spectrum of design activities between the areas of design focused largely on form – such as one might consider with the design of clothing or furniture – and engineering design which focuses primarily on function. Somewhere in the fuzzy boundary between the two lies industrial design, the career that put Jonathan Ive in the position of dictating both the form and functionality of some of the most popular gadgets of the 1990s and 2000s. In his role within Industrial Design at Apple, Ive revolutionised the appearance of personal computers and personal electronic devices like music players and phones.

A product designer, Ive studied industrial design before working in a series of design companies, finally settling with Apple. Does the work that Ive does bear any relation to engineering? It

JONATHAN IVE AND APPLE (*cont.*)

certainly pushes against the boundary of engineering. For a start, an essential aspect of the work of an industrial designer is the consideration of how products will be made. It is this focus on manufacturing process that introduces engineering into the world of the designer, bringing into the design process considerations of the cost, efficiency and organisation of the manufacturing process.

This concern with the manufacturing process is central to the work of Ive and others at Apple in using standard materials in innovative ways to produce novel effects. A profile of Ive in Business Week commented thus on Ive and Apple's work with plastics: 'Take Apple's pioneering work in injection molding. It's part science, part art, and plenty of trial and error.'[5] This is almost a definition of engineering.

There is no real gain in trying to argue that Ive is an engineer and there is no doubt that his role is quite different to that of an electronics engineer who designs and develops the hard disk that stores the files on an iPod music player. But his work illustrates that the apparently diverse worlds of engineering and aesthetics do rub up against each other, and that the engineer is related to the artist as well as to the scientist.

Engineering and design are also closely related concepts, but design encompasses a range of activities of which engineering lies at one end of the spectrum. The engineer devises ways to create things which do not yet exist but will fulfil some need, and as such is a designer. This activity distinguishes the engineer from the scientist who sticks to what already exists, and the trade or crafts person who makes what they are asked to, according to plans and instructions provided for them. However, design, in popular understanding, is not identified with engineering. Design is very often associated with aesthetics and the archetypal designer in most peoples' minds designs clothes, furniture or book covers.

In engineering design, function is always of far greater importance than form or appearance. Engineering design involves scientific analysis as well as creative skill, because the design process is about exploring the physically possible means of creating an artifact to fulfil some need. Increasingly, this is aided by computer simulation to explore and refine different design solutions, and is a far cry from the freely creative sketching of a clothing designer. But the centrality of design in engineering means that engineering is essentially a creative discipline, involving dreaming up solutions and trial and error testing. This aspect of engineering is often downplayed in favour of an emphasis on the application of scientific theory and mathematical rigour in engineering.

Engineering is a complex discipline. It took many centuries to emerge, and it shares borders with many other disciplines. This is unsurprising giving that it is so essentially intertwined with our lives – it touches many aspects of the way we live, and has undergone developments in line with, and often in support of, leaps forward in quality of life. The history of engineering plays a large part in human history. The next chapter will outline the specific ways in which each engineering discipline has ushered in new phases in human history.

2

The world of the engineer and the engineering of the world

The history of engineering is a history of diversification and specialisation, as the creation of new technologies and the emergence of new skills and techniques spawns engineering sub-disciplines with expertise fit for the niches created. The evolution of engineering into diverse species began with the division of civil from military engineering, which was followed by the mechanical, electrical and other engineering groups distinguishing themselves from the kinds of projects engaged in by the original civil engineers. In addition, engineering specialisms were created by existing crafts and trades developing into engineering disciplines through the introduction of engineering method.

This chapter will set out the ways in which engineering has changed the physical world. It will show what each engineering discipline involves by examining the impact that each has had on the world around us and the way that we live in it. But it should be recognised that each individual discipline of engineering can have only limited impact on its own. The lines between the specialisms are vague. Many engineers sit on the border of a variety of specialisms and most significant engineering projects are a collaboration between a number of disciplines. One of the essential skills of any breed of engineer is to be multidisciplinary, to be able to work with experts from other areas and to see where their work fits into the wider scheme. Engineers as a collective have changed the world, and no single discipline can claim to be the most influential.

Civil engineering: the creation of cities and the taming of the country

As the last chapter demonstrated, civil engineering is the oldest branch of the discipline that we think of as modern engineering. The 'civil' in civil engineering may initially have distinguished those engineers who worked on non-military projects from those that supported the work of soldiers at war, but it also connects intimately to the concept of civilisation. Living a healthy and safe life in large settlements and cities is made possible by civil engineering primarily through its management of water, the key element in supporting human life. Thus civil engineers plan, design and build the hidden labyrinth of pipework that allows the distribution of water to homes for drinking and washing but, perhaps more importantly, they construct the tunnels through which wastewater and sewage is flushed away. The management of clean and contaminated water makes life possible in large towns and cities, and without it we would be, and indeed once were, at risk of poisoning from our own waste. Civil engineers also protect us from the threats that water poses, by building flood defences and dams – allowing us to tame the forces of nature which formerly controlled the place of water in our life.

Another key element of modern civilisation is the transport system. The transport system allows us to work away from where we live, in centralised offices and factories, and allows us to maintain contact with those we live away from, by allowing cheap and fast travel. Civil engineers produce and maintain the transport infrastructure; the road and rail networks with their bridges and tunnels that allow transport systems to overcome natural barriers. The transport system is not only important for allowing individuals to move around but is also of course essential to industry by making possible the movement of goods produced in one part of the country to another, and also to the shores for export. Before the industrial revolution and the rapid expansion of the railways, civil engineers shaped the waterways of the countryside so that they could meet the transportation needs of mining and industry. The transport infrastructure was also at one time the communications infrastructure – roads and rails making it possible for a postal service to exist, which in turn supported modern industry and personal communications.

The civil engineer is not however the person who gets their hands dirty digging tunnels or laying tarmac. Civil engineers are project managers, managing huge taskforces that are essential to carrying out large scale projects like the creation of a motorway or a sewers network. This aspect of civil engineering connects the modern engineer to the early proto-engineering work of the Romans and Egyptians. The building of the pyramids no doubt involved some of the key skills in civil engineering, in terms of the management of huge labour teams and the movement of construction materials. Those skills are more sophisticated now, with civil engineers using modern methods and computing software for managing construction projects. They also employ mathematical methods to ensure that the tunnels and bridges they design will not crack or shift, and to analyse the nature of the land in which they are building in order to inform the design of their projects. Civil engineers have had to develop new skills for ensuring that their work is sustainable, and that it will not have an adverse impact on the environment.

The great civil engineering projects might seem to be over – we have the infrastructure we need, it simply requires maintenance. But there are parts of the world still without clean water and sewerage systems, and while these exist there will always be a challenge for civil engineers. And the maintenance of our infrastructure is, even in developed countries, a massive and critical task. The general growth of the global population and the constant swelling of the world's largest cities means that the transport and water infrastructure is under constant pressure from increasing demand. Many cities depend on water and sewage systems created generations ago, which are under threat of decay but so deeply embedded in our towns and cities that renovation is a disruptive and costly project. As long as there are people, there will be civil engineering challenges.

It could be argued that it is the civil engineer who has had the most profound human effect on the landscape of our world, and some of the most fundamental changes have been as a result of civil engineers influencing the place of natural water in our lives. Take, for example, the Hoover Dam. Built to stem the flow of the Colorado River in order to irrigate the desert land that surrounded it, and to harness its power for the generation of electricity, it was one of the most audacious attempts to change the face of the landscape. Completed in 1935, and creating a man-made lake 150 miles long, the dam is a sheer, human-made concrete cliff rising 727 feet from the river bed between the walls of Black Canyon, 30 miles from Las Vegas. The dam changed the landscape in which it was situated from desert to habitable land nourished by the Colorado, and constitutes in itself an imposing addition to the face of the country. Civil engineers are responsible for some rather more subtle interventions, though. Telford's Caledonian Canal utilised the natural lochs traversing the belt of Scotland and extended them to create a continuous waterway. This was a project repeated in kind but at far greater scale by the creation of the Panama canal, which dissects an entire continent, creating a navigable shipping path through America.

Are such things seen as impositions on the landscape? Do we, however reliant on them we might be, see the results of civil engineering as changing the environment for the worse? Many canals and reservoirs have almost become part of the natural landscape and people enjoy them – across Europe people take cycle trips along canal ways, which is rather like picnicking by the motorway if you consider their initial purpose. Are they acceptable because they are established and have become part of our heritage? Or are they accepted because, although they have changed the face of the countryside, they have not destroyed it? In many ways, these changes are a continuation of the changes that agriculture has made to the landscape. Civil engineers were not the first to change the way that the natural landscape looks through changing the way that we live in it. Huge parts of the rural world are radically different to the way that nature made them because of the way farming has sculpted them to be fit for purpose. Is the impact on the natural landscape made by civil engineers similar in kind to the changes made through the development of agriculture? These are no doubt questions that civil engineers ask themselves, as they become ever more concerned about the impact of their work on the environment.

Structural engineering: taller, lighter, stronger

Civil and structural engineering are closely allied professions, in that both elevate what might have been simple building work to the level of large-scale, ambitious and organised construction projects. The subject matter of structural engineering is obvious – the design and construction of structures – but its methods are its uniquely identifying factor. Structural engineering is about applying analytical methods to the design of structures in order to be able to build larger, lighter constructions that are safe and resilient. It is also about the development of architectural materials that have the right properties – strength, lightness, flexibility, and reasonable cost – to realise the ambitions of architects.

There is by no means a structural engineer behind every building – architects and builders can often manage without the direct input of the structural engineer. This is in part because the knowledge of structural engineers is distilled in text books and building regulations which are used and followed in building design and construction. However, modern skyscrapers, stadiums, railway stations, airports, hospitals and other large and complex buildings depend for their existence on the active involvement of structural engineers. Bridges have been created to span ever longer distances whilst exhibiting delicacy of form, requiring processes of structural analysis to underpin their design. Complex vehicles such as ships, trains and aircraft need the input of structural engineering. Frame engineers are employed to design the structure of large vehicles and vessels which will be subject to stresses induced by operating at high speeds under the destructive influence of the elements. Structural engineers thus work closely with a number of other disciplines – not just architecture and civil engineering, but also mechanical and electrical engineering.

The structural engineer needs to understand how to develop structures so that they will (literally) stand up to both normal and extreme conditions, and crucial to this is understanding why structures fall down. Since a failure in structural engineering can result in the loss of hundreds or thousands of lives, understanding structural failure is a hugely important task. Building and bridge collapses have been studied and analysed in depth to devise ways to create structures that are less susceptible to failure. For example, studies of the way that the Twin Towers collapsed in the tragic events of 9/11 have been used to devise methods for designing tall structures that are more resilient to the effects of fire.[1] Structural engineers carry a weighty responsibility to ensure that the buildings they work on are robust enough to withstand greater loads and forces than they are ever expected to experience, and this involves ensuring that the right materials are used and that construction is carried out exactly as prescribed.

Structural engineering has thus made possible some of the most salient and striking non-natural landmarks in the world, and has changed the face of modern cities. The Shanghai Pearl, the London Eye and skyscrapers from the Empire State to Taipei 101 owe their existence to the growth in understanding of structures. Gustave Eiffel can be singled out for two of the most famous feats of structural engineering – creating the Eiffel tower and collaborating on the design of the internal scaffold which forms the sturdy skeleton of the Statue of Liberty. But there are armies of often unrecognised engineers behind the buildings that dominate large cities. Skyscrapers are only possible by virtue of a revolution in understanding of how to build high-rise structures. The Crystal Palace, built in Britain for the Great Exhibition of 1851, is the perhaps unlikely inspiration for the modern building due to its 'curtain wall' method of construction. The iron and glass structure was built by creating a supporting metal frame, from which glass walls were 'hung'. This was a diversion from existing methods of construction, where the higher floors of a building were supported by the walls underneath them.[2] The new method of construction made high rise structures possible, as buildings could now be built to significant heights without having to make the masonry walls of a building extremely thick to support the higher floors. Cities such as New York and Tokyo are in many respects showcases for the contribution of structural engineering in sculpting the face of the modern city.

Many of the most notable structural engineering projects are achieved through a sometimes painful process of collaboration

OVE ARUP AND THE OPERA HOUSE

Few structural engineers are as high profile as the architects that they collaborate with, but Ove Arup (1895–1988) arguably ranks as one of the best known, largely through the works of his eponymous company of consulting engineers. The Sydney Opera House was one of Arup's highest profile projects.

The original architect of the Opera House was Jørn Utzon, whose diagrammatic sketch of a building captured the imagination of the panel judging an open competition for designs. Arup (the company) was appointed as the structural engineer for the project. The central feature of the Opera House is its roof, made from a series of shell-shaped concrete vaults. The irregular shape of the vaults presented a significant engineering challenge, and a central task for Arup was to build a model of the roof, and subject it to pressure testing to assess its resilience to wind. The complexity of the design demanded a number of changes throughout the process of design and construction, and political pressure to complete the project meant that some parts of the Opera House were constructed before plans for other parts were finalised. This left Arup engineers with the task of adapting to design changes after much work was already completed.

The tensions between the architect, building contractors and politicians made this a tortuous task for the engineers. Ove Arup said of the project: 'I would say we are quite crazy to take this job. I am not complaining, I'm just saying how hard it is. I never realised it would be so hard... It's taken two or three years off my life.'[3] He was not the only protagonist in the story to suffer. Eventually, as a result of disagreements with the government, Utzon resigned as architect.

The Sydney Opera House was pushed into life without clear ideas about the construction process, timescale or budget. The first estimate of cost was given as $7m in January 1957; the final figure was a staggeringly disproportionate $99.5M. The completion date skidded by almost a decade, and it was officially opened by the Queen in 1973. However, despite the tribulations that Arup suffered during the work on Sydney Opera House, it was an undoubted success for them, as they pushed the boundaries of engineering to become midwives of one of the world's most breathtaking and recognisable human-made landmarks.

between architects, engineers, construction firms and, in the case of public buildings, politicians. The work of the Arup company on the Sydney Opera House (see box) is testament to the tribulations that engineers can face. However, tools have been developed that are making the process of creating complex buildings ever easier. Computer-based design tools allow plans to be made and shared rapidly between large teams based in different parts of the world, speaking different languages. The distilled understanding of generations of structural engineers are embedded in design packages that allow engineers to create ambitious designs but with greater control over the risks involved.

Mechanical engineering: from iron horses to 1000 brake horsepower

Structure and infrastructure are at the heart of engineering but its soul lies, surely, with the machine. Just as the words 'engine' and 'engineer' are linked by etymology, in the imagination of the average person the engineer is one who dreams up, constructs, tinkers with and mends engines. From designing to fine-tuning the engines of racing cars, the 'engineer' and the 'mechanic' in mechanical engineering are closely related. Mechanical engineering forms a spectrum of activities, merging with manual crafts at one extreme and physical sciences at the other.

The core pursuit of mechanical engineering is harnessing and using natural power and converting natural fuels into useful work. Thus mechanical engineering has roots deep in history, stretching back as long as humans have created machines to extend human strength and capacity for work. The power of water, wind and the strength of animals were the first to be exploited, followed by the power of steam in the time of the industrial revolution, through to the energy created by a nuclear chain reaction in the twentieth century. Wind and wave power are again becoming important, but now they are harnessed to generate electricity.

A crucial point in the history of mechanical engineering was the invention of the steam engine, based on the steam pump designed to extract water from deep mines. The steam engine developed rapidly as the demand for coal grew during the time of the industrial revolution, and significant effort was focused on its improvement. This process continued for many decades, and the significant progress made in steam engine design and construction was central to the first transport revolution, as it facilitated the creation of the steam locomotive and the paddle ship, and even steam powered cars. The internal combustion engine was developed at the end of the nineteenth century and had some employment in industry, but the real impact of this technology was in its use in the motor car, which sparked the automobile revolution. Mechanical engineers were also at the heart of the electrical revolution, as mechanical engineers are key to the design of coal-fired, nuclear, hydro-electric and all other breeds of power stations.

The need for power is complemented within mechanical engineering by the need for precision. Neither the steam engine nor the combustion engine would be possible without the skills for producing precision components to make the engines. The process of machining, producing the metal components for machines in a uniform and precise way, is a crucial part of mechanical engineering that continues to develop as ever more precise components are needed for increasingly sophisticated machines.

The development of the products of mechanical engineering created their own sciences, due to the need to better understand the laws that govern the functioning of engines and the vehicles that they power. Thermodynamics arose out of an attempt to understand the movement of heat and power through a system such as a steam engine. While engineers such as Thomas Newcomen and James Watt in the UK were improving the function of steam engines, the French scientist Sadi Carnot was analysing the principles by which they work, and his work formed the basis of the second law of thermodynamics. Fluid dynamics and aerodynamics, which describe the movement of bodies through air and liquids, such as ships or aircraft, or the movement of fluids through machines such as steam turbines, are important tools for mechanical engineers. They can be used to design more efficient engines and vehicles such as steam turbines and aircraft. Other tools essential for mechanical engineers include mechanics of materials and stress analysis to design machines and structures that are strong enough, rigid enough and durable enough for their intended purpose. The mechanical engineer has a sophisticated grasp of these applied sciences and their use in a range of design problems from the improvement of a car engine to the architecture of a complete warship.

The impact of mechanical engineering on the way that the modern world appears and functions is immeasurable. The development of the steam train gave rise to 'railway mania', the frenetic investment in and construction of railways across the UK, which was mirrored throughout the world. Railways shrank the countryside, making long distance travel and transport possible. The effect of the railway on the US was particularly dramatic, as the huge engineering project of the transcontinental railway connected the coasts of the continent and made a path through uncharted and untamed territory. The vast railway system of India had a similar impact, and is still a crucial part of the transport infrastructure. The motor car too has changed the lives of many people across the world, from Ford's mass produced Model T to the 'Nano' car developed by Tata for the Indian market in 2008, costing just over £1000/$2000 at its launch. And of course air travel has made our world ever smaller, enabling us to cross the world in a day and making overseas travel an option for increasing numbers of people. There is not a part of our lives that is unaffected by the work of mechanical engineers in transport. Many countries are dependent on transport to bring food and fuel from its source to the consumer. When transport systems are affected by accidents or extreme weather, chaos quickly ensues.

Mechanical engineers are involved in the design of machines that are both precise and powerful, and they have a hand in some of the biggest challenges in engineering, from creating functional prosthetic limbs to developing sustainable energy solutions. The vast majority of means for generating electricity from renewable and low-carbon resources, wind and wave power in particular, will need the skills of mechanical engineers. Mechanical engineers will also have an important role in ensuring that our most energy hungry technologies, from aircraft, to sports cars, to air conditioning systems, are more fuel efficient and thus less environmentally damaging.

THE JAMES WEBB SPACE TELESCOPE

Satellite engineering is a multidisciplinary area – encompassing mechanical and electrical engineering, and heavily dependent on software development – which has expanded the boundaries of the experienced world. Observation of space by satellite has been crucial to developing our understanding of the universe by allowing us to see deeper into space, and thus further back into the history of the universe. The James Webb Space Telescope (JWST), designed to replace NASA's Hubble Telescope, is the creation of an international team of engineers and space scientists. Hubble (and results from observatories) revealed that a great

THE JAMES WEBB SPACE TELESCOPE (*cont.*)

deal can be learned from observations made in the infra-red portion of the electromagnetic spectrum. It is at this end of the spectrum that bodies further away, and much further back in the history of the cosmos, can be observed. The JSWT will host MIRI – the Mid-Infrared Instrument – which will seek out light from the first stars formed after the big bang. It will also observe the formation of new stars and galaxies by seeing through the dust that surrounds new-born stars which is transparent to infra-red radiation, but opaque to visible light. Because it is seeking out infra-red light, MIRI must be kept extremely cool so that the telescope itself does not radiate heat, causing interference. Parts of the telescope need to be less than 6.7 degrees above absolute zero to function.

There are phenomenal engineering challenges involved in such a project. Just testing the satellite's cooling systems involved building test facilities that simulate the frigid environment in which the satellite and cameras will function. A project like this is a highly complex piece of engineering not only in the technologies needed, but also in the coordination of teams of engineers and scientists working on each intricate part of the whole in 12 different countries. Enlightening us as to the origins of our world is no longer a matter for the isolated reflection of the ancient philosopher, but a technology-driven team effort.

Satellite engineering is not all about looking deep into space, as a huge application area for satellite engineering is earth observation. The first pictures of the earth from space shaped the perception of the world for subsequent generations. Now, satellites routinely take images from space for a variety of everyday purposes such as tracking traffic through GPS (Global Positioning System) and monitoring the development of crops so they can be harvested at the optimal time.

Figure 1 Fully functional, 1/6th scale model of the James Webb Space Telescope mirror in optics testbed. *NASA*

Electrical engineering and electronics: the democratisation of power

Electrical engineers design and create devices powered by electricity and the systems that allow the transmission and distribution of electrical power which supplies those devices. The development of electrical engineering was heavily dependent on discoveries within physics, where the phenomena of electromagnetism was being studied by key figures such as Michael Faraday and James Clerk Maxwell. What was initially phenomenon of interest only to scientists was exploited by engineers who saw the potential in generating and distributing electrical power.

One of the drivers for the distribution of electrical power was the movement towards the wide-spread installation of electric lighting. Central to this process was Thomas Edison, a figure impossible to avoid in any story about the development of engineering. Electrical light was not so swift and sudden a revolution as might be thought, and people were not necessarily keen to replace gaslight with electric light, imagining electricity to be potentially dangerous and unreliable. But this initial resistance was soon overcome and for billions of people life is unimaginable without access to 24 hour power and light. The march of electrification is not complete however – it is worth considering that the number of people living without access to mains electricity is usually estimated at 2 billion.

The progress of electric light did not stop with Edison's attempts to perfect the lightbulb, but continually grows, with the development of sophisticated lighting systems embedded in modern buildings and novel means of creating light to replace the energy hungry filament bulb. From there we have moved through fluorescent lighting and on to solid state lighting – such as LED lights – which hold the promise of cheaper, long lasting and energy efficient light sources.

The draw of electric power brings engineers from other disciplines into the arena of electrical engineering, as electricity offers a clean and potentially low-polluting form of energy (depending, of course, on how it is generated). Just as the steam train was replaced by the diesel train, the electric train now dominates, and an engineer working on the development of railways is as likely to be an electrical as a mechanical engineer.

Similarly, the demand for electric cars has grown with the need to limit pollution, and the development of hybrid and electric cars for market has been an important area of work for electrical engineers.

It doesn't take much imagination to consider just how central electrical engineering is to our experience of living in the world. Alongside civil, structural and mechanical engineering it helped to create the very essence of the modern city. According to the historian Thomas Hughes, 'electricity made modern New York possible'. Lighting, power, elevators, electric trains, electric motors for industry, all are essential aspects of town and city life. Without electrical power to operate elevators, the masterpieces of structural engineering like the Eiffel Tower or the Empire State building would be white elephants, as only the athletic would get to see the view from the top. Electrical engineering is integral to the visions of structural engineers, providing their beating hearts, giving them heat, light and power.

Electrical engineers develop a solid grounding in power generation, transmission and motors. The scientific foundation of electrical engineering lies in an understanding of electromagnetism and of physical laws such as Gauss's Law (which relates electric charge to the electric field) and Fleming's left and right hand rules (which describe the relationships between the directions of motion, the electrical field and current flow in motors and generators). Electrical engineers become expert in the design and analysis of circuits, making use of Kirchoff's laws, which describe the conservation of charge and energy in a circuit.

As with mechanical engineering, electrical engineering has rapidly moved forward and created new sub-disciplines as the technologies it gives birth to become more specialised. An area closely allied to electrical engineering is electronics, the discipline of understanding the behaviour of electrons and using their properties to develop electronic devices. Electronics is where engineering and physics merge – it is hard to draw a line between electronics done in the engineering department and the physics laboratory and many engineers working in this field will deny any clear distinction between the work that they do and that of a physicist. The development of semiconductors, for example, lies at the intersection of physics and engineering. Semiconductors are materials which are part way between electrical insulators and conductors (materials through which electricity can flow), and a major use of semiconductors is in transistors. Transistors are tiny devices that have applications in nearly all electronic devices from radios to computers.

The electronic engineer will have a deep understanding of Maxwell's laws governing electromagnetism, essential for understanding the behaviour of radio waves in telecommunications. Semiconductor physics is essential for working in the area of integrated circuits (otherwise known as microchips). Like many other areas of engineering, electronics makes use of software tools for the design of circuits and electronic devices with a number of proprietary software products being used by students and professional engineers. As electronics and electrical engineering has developed it has been able to supply itself with ever more sophisticated tools. Analogue devices for measuring voltage and current, or spectrum analysers which study the frequency components of signals have been replaced by digital counterparts which display data on digital displays rather than analogue dials which the engineer would interpret.

Electronics may currently be in the midst of a new revolution through the development of plastic electronics, a growth area where physics and engineering are working together to produce life-changing technologies. While electric circuits could previously only be constructed on rigid surfaces such as silicon, plastic electronics allows circuits to be printed onto flexible structures. This means that electronic gadgets can be lighter and more pliable, and flexible, rollable displays can be made for mobile phones and electronic reading devices. It even gives the promise of creating medical dressings with embedded electronics, creating 'smart bandages' that can be designed to monitor and treat wounds while they heal undisturbed. Plastic electronics might even be the key to developing photovoltaics much more cheaply. Photovoltaics is about harnessing natural light to create electricity (a process first described by Einstein, in the work that won him his Nobel prize) but which, despite the use of solar panels for small scale electrical generation, still does not generate electricity in a cheap and efficient way. The use of plastic electronics to manufacture solar cells could make this process much cheaper and more accessible. So, while one of the first challenges for electrical engineering was to turn electricity into light, there is a continuing challenge to turn light into electricity.

THOMAS EDISON – AN ENGINEERING DYNAMO

Edison was an innovator and inventor of extraordinary tenacity; his breakthroughs were not insights that arrived with the flick of a switch but hard-won developments that took time and team effort. Edison set up his famous Menlo Park laboratory in New Jersey where his team was set the demanding target of delivering a minor invention every ten days and a major one every six months. By working in such a way Edison produced the phonograph – a device for recording the human voice by making an impression on waxed paper – a precursor to the gramophone that wowed the world and earned him the title 'the Wizard of Menlo Park'.

Edison's name will forever be associated with the light bulb. He was not the inventor of the filament light bulb, but was an important improver of the technology, spending many hours searching for the best material for the glowing filament. Edison's

THOMAS EDISON – AN ENGINEERING DYNAMO (*cont.*)

concentration was set on the goal of devising a complete lighting system – encompassing the production of electricity and its distribution to homes and businesses along with the provision of lighting devices. The development of such a system was quite groundbreaking and risky; Edison said of his system: 'What might happen on turning a big current into the conductors under the streets of New York, no one could say.'

Edison's inventiveness sprang up in a great business mind meaning that his innovations were focused on mass market and commercial potential. His intent on business success also meant that he became embroiled in various court cases and somewhat undignified battles. He based his lighting system on direct current (DC) – largely because the advantages of alternating current (AC), which could be distributed over long distances, were initially offset by the lack of an effective AC motor. However, the Serbian-American inventor Nikola Tesla designed such a motor, encouraging the American electrical company Westinghouse to invest in AC. This new system put Edison's electrical distribution system under great threat, and he did all he could to prove the inadequacy and even deathliness of AC. Despite this, AC won out as the better system. It is open to question whether Edison's dogged support for his system was mistaken pride, an interest in personal success over the growth of technology, or simply a necessary mindset for someone with the will to push through barriers and to forge new technologies.

Information and communications technology: engineering a new revolution

ICT justifies talk about revolutions because there is no doubt that the world of the networked personal computer is radically different to the world before it. Just as industrial machinery fundamentally changed the way that working people lived out their days – moving from cottage industry to factory, working away from the home in industrial centres – the computer has changed the working lives of many, creating new jobs and making others obsolete. It allows people to work away from the industrial centres in their own homes, connected by the Internet and email, and allows cheap and fast communication with colleagues worldwide. None of this would be possible if engineering ingenuity had not been focused on making smaller, cheaper yet more powerful computers.

Mechanical and civil engineering made the world smaller by making physical links between distant places, but information and communications technologies have shrunk the world almost to a point, making it possible to communicate with other people, send documents, and engage in financial transactions almost instantaneously. Conversely, ICT has expanded the world massively by creating a whole virtual annex to the 'real' world. In developed countries people of all ages spend hours of their day using their computers to buy products and services like clothes and holidays in online stores; to communicate with friends and family via email, instant messaging and social networking sites; and to watch the news, films and television whenever they like. A living can be earned within the virtual world by making and selling items that can be used in multi-player online games, with users in poorer countries making money by slaying dragons for time-poor gaming fans in richer countries.

As with many engineering-led changes, the computer revolution had a long gestation period, with disruptive technologies appearing some time before they really impacted on the lives of the masses. Queen Elizabeth sent her first email on 26 March 1976 (the occasion was the official formation of the Royal Signals and Radar Establishment in the UK and the email utilised the connectivity provided by ARPA for the Defence community – see box on p. 46). One of the main factors in this lag was the wait for computers to be cheap enough to be bought by significant numbers of businesses and individuals. It was only when ownership reached a critical mass that technologies like email and the Internet became necessary for everyone. A huge factor in this was the development of transistors, discussed above, which were small, cheap and easy to manufacture, and thereby led to smaller and cheaper computers. But developments in software were also crucial to the popularity of the computer, from the proprietary word processing packages that made the computer a more worthwhile desk companion than the electric typewriter, to the development of the World Wide Web and the web browsers that allow non-technical computer users to access the Internet. The result of this permeation into the mainstream was the rapid and full-blown communications and computing revolution that we are still living in. Now that it has taken off, innovations in the networked world of information and communications technology happen at lightning speed – the online community SecondLife grew rapidly after its launch in 2003 to having the equivalent of a gross domestic product (GDP), based on its currency Linden dollars, larger than most countries in the world.

While this infiltration of ICT into all areas of life may make it seem that we are becoming slaves to computers, the landscape is continually developing so that we can enjoy connectedness while being physically free and untethered. Mobile computing is in continual development and allows access to the Internet on the move, with increasingly powerful hand-held devices delivering what could once only be accessed at a desk. The end goal is access to all information – from the whole of the Internet to your own personal documents – wherever you are and whenever you want. ICT has also been a great enabler in the developing world, with mobile telephones allowing communications in remote regions where establishing fixed telephone lines would be unfeasible. Focused efforts were made to develop cheaper and more robust laptops which would be useful to children studying in parts of the world where funding for education is limited and the classroom spills into the outside world. But this growth brings with it some serious downsides – the consumption of power for computing is rising dramatically and the carbon footprint of our addiction to connectivity is outgrowing that of aviation. The challenge to develop lower-energy computing is one of the next hurdles for ICT.

So what is the engineer's role in all of this? Although their obvious impact is via electronics and the development of computer hardware, some would argue that engineers have contributed just as much to software. 'Software engineering' and 'software engineer' are phrases that have some currency, suggesting that the development of computer programmes and packages are the product of engineering. But many software developers see little engineering in what they do. Software designers do not need engineering degrees and simple systems can be developed with minimal application of mathematics or basic science. However, software is now part of the critical infrastructure of most of the world – it supports banking, retail, healthcare and allows the better management of utilities like power and water. An IT failure, as everyone will have experienced at some point, can bring non-IT services to a grinding halt. Therefore, software could do with some of the skills of engineers who have experience in delivering critical infrastructure. And in fact, it is the centrality of the user to software that makes engineering especially relevant. Anything an engineer develops must meet a pre-existing need or stimulate a new demand, so engineers must understand how best to meet the needs and desires of users. This is as relevant in IT as it is in building design. Indeed, some argue that the reason for talking about software engineering at all is in the hope that software development will become like a branch of engineering, and will become all the more rigorous and reliable for that.

To shoulder this responsibility adequately, software engineers need to develop skills in computational thinking, using logic, algorithms, and methods of abstraction to represent and analyse the systems that they will create. This enables them to design robust systems with rigorous foundations. They will learn to use notations like UML (unified modelling language) and programming languages such as SQL (structured query language) and Java. But a software engineer also has to link the system that they are developing to the needs of users, so the ability to understand a client's needs and articulate them in clear, unambiguous everyday language can be as important as mastering formalism.

TIM BERNERS-LEE, HTML AND THE WEB

Despite many people's assumption that they are the same thing, the Internet predates the World Wide Web by some years. The Internet itself was preceded by the Arpanet, created for the Advanced Research Projects Agency (ARPA) in the US, which was commissioned in 1969. The Arpanet established a network of computers in different US universities; the Internet was about establishing a network of such networks, which eventually stretched worldwide.

The early Internet allowed the transmission of emails between computers, and it allowed computers to view documents stored on remote machines. But there was a lack of uniformity in the way that different computers stored information and communicated it. Different 'protocols' (the means by which a computer breaks down data to be sent out on the Internet and puts received data back together) existed for different networks, blocking the possibility of a global network. The vision of Tim Berners-Lee was of complete universality: anyone should be able to access anything that is available on the Internet on any machine using any software, however it is connected to the Internet.

TIM BERNERS-LEE, HTML AND THE WEB *(cont.)*

The three elements of the web are URLs, the Uniform Resource Locators which constitute the unique addresses for everything on the web; HTTP, Hypertext Transfer Protocol, the set of rules by which computers communicate with each other to share documents; and HTML, Hypertext Mark-up Language, which describes how a webpage should look on the screen. The system of universal addresses is the most fundamental concept of the web. It allows any computer to find any document that is on the web, and allows the linking of one document to another by means of hyperlinks. It is by virtue of these links that the World Wide Web is truly a web, with any document being able to connect directly to another.

Despite the enormous impact of the World Wide Web, Berners-Lee made no personal profit from it. His aim was rather to produce a system that was not owned by anyone, and in which any user could put anything that they liked on the Internet – it was a forum for communication and collaboration. This vision came to be a little later than the Web itself, coming into its own with the development of user-generated content, such as blogs, discussion forums and social networking sites.

Chemical engineering, manufactured materials and nanotechnology

Engineers have not only changed the world by building on it and connecting it through transport links, they have adapted and added to the very stuff that it is made of. Chemical engineering and materials science are disciplines that lie on the border of science, getting to the very building blocks of the stuff that nature has produced, and improving on that found matter and making it more suitable to human needs. Our world differs fundamentally from that of our ancestors in that many of the things we surround ourselves with and use everyday are made of materials they would not have dreamed of, allowing for functions that they might have only dimly wished for. The development of materials for human use is not a new thing, but weaves through all of human history. Indeed, we are only now learning the secrets behind the tricks that our distant relatives could do with materials; but through that engineering knowledge we can turn those tricks to our greater advantage.

Chemical engineering is the discipline that turns raw materials into substances that meet some need, many of which are critical to our everyday lives. For example, the petrol powering cars and public transport is not a natural product that is pumped straight from the oil well to the petrol station forecourt, but is a manufactured product that has to be refined before it is usable. Chemical engineers are able to do this because they have taken the understanding of the basic structure of substances that chemists have developed, and the kind of lab-based experimental processes that chemists employ, and scaled these up to produce and process chemicals at an industrial scale. The essential skill of a chemical engineer is that of being able to 'scale up' reactions that are creatable and controllable in the lab so that they can be harnessed and employed to create the substances we need in the quantities that we demand. This involves not only a scientific understanding of the nature of the substances being processed, but knowledge of the methods needed to ensure that volatile chemicals are processed in a way that minimises the significant risks that they pose when handled in large quantities. This is no trivial matter, as these volatile substances may behave in different and unpredictable ways when handled in very large quantities.

In an age when we no longer draw water from the well or spring, the filtration and cleansing of water so that it is safe to drink is an industrial process that is designed in part by chemical engineers. This might seem to be a skill that was crafted many decades ago, but the challenge of turning naturally found water into something potable continues to be a challenge in developing countries, where the aim is to find ways of producing safe and clean drinking water where disease is rife but infrastructure minimal. Where huge civil engineering projects to create sewers and pipe in clean water are impossible, chemical engineering has a great deal to contribute in this area, in terms of finding ever cheaper and more accessible ways of making water safe for remote or impoverished communities.

Our food is increasingly the product of large scale industrial processes that are designed by engineers. Many people in the world have diets augmented by foods that have been made in large quantities, to a specific standard and which will have a long shelf life. From emulsifiers in low fat mayonnaise to the flavouring in sour candies, our diets have been fundamentally changed by engineering, for good or bad. The same is true for healthcare – the manufacture of medicines has moved from the apothecary with a pestle and mortar to massive pharmaceutical plants which produce drugs in quantities that will serve a global market. The particular challenge here is producing drugs in huge quantities whilst ensuring that every tablet or capsule that is delivered contains exactly the right amount of active ingredient – even miniscule variations could have health-threatening consequences. Without chemical engineering the mass availability to large populations of the world of safe and relatively cheap drugs would not have been achieved. As these processes are developed, there is hope for cheaper drugs for some of the world's most challenging diseases, affecting some of the world's poorest communities. Many chemical engineers currently work in the petrochemical industry, supporting the world's huge appetite for petrol but they will no doubt be central to the development of alternatives to petrol, including the development of hydrogen fuel cells which are seen by many as the clean alternative to fossil fuels.

One of the core concepts that a chemical engineer has to grasp is that of mass transfer – how one substance changes to another, at what rate and in what conditions. A physical process as simple as water boiling involves water in a container evaporating and moving into the surrounding air – the chemical engineer will have to grasp just this kind of change but for a host of far more complex processes. This sort of concept is important, for example, in the development of drugs, as understanding how and at what rates a drug delivered in tablet form will be released into and absorbed by the body is important for producing drugs that can deliver low doses over sustained periods of time. The chemical engineer will learn about material and energy balancing and auditing – understanding the relationships between the material input and output of a process and the energy used by it. The methods that allow a chemical engineer to control these processes include mathematical modelling of these processes so that they can be understood and anlaysed in the abstract, and project management and process flowsheeting in order to manage and control the real life processes. Chemical engineers will make significant use of software for modelling systems and for running simulations of the processes that they will design and implement.

Closely allied to chemical engineering is the area of materials science. Whilst a large part of chemical engineering is about refining what nature provides, chemical engineering and materials science are also in the business of adding to the stock of the types of material in the world. A list of human-made materials that are the products of science applied through industry would be near limitless – as would a list of all the crucial things in our life that are made possible by invented materials. Materials scientists and engineers familiarise themselves with the molecular and atomic structures of matter to produce materials that have desired properties at the macroscopic level – being light or fluorescent, strong or stain resistant, malleable or anti-bacterial. Materials are engineered to stop fried eggs sticking to pans (Teflon); to make lighter vehicles from racing bikes to more energy-efficient aircraft (carbon composites); to improve human performance in sport (Speedo's LZR Racer swimsuit that helped to break multiple world records in the 2008 Olympics); and to fix broken body parts (hard-wearing and biocompatible materials in prosthetics like hip joints). The challenge for engineers is to find a way to make materials with the desired properties, that can actually function in application, and which can be manufactured safely and at reasonable cost to make commercially viable products.

It is not only wholly novel materials that promise revolutionary applications. Nanotechnology is an area of materials science that exploits properties of materials that emerge at the very small scale, which can be quite different to the properties of those same materials at the large scale. The scale of interest is between 100 and 0.2 nanometres – where a nanometere is one billionth of a metre; natural objects that fall within this range are viruses and strands of DNA. When materials are at these miniscule scales they behave quite differently, due to the quantum mechanical effects that appear in the microscopic world. So, for example, nano-sized particles of gold can have quite different colours, such as red or blue, and in fact this feature of nano-sized particles has (unwittingly) been exploited in making stained glass for centuries. We have only recently learnt how these effects were possible.

Nanotechnology is all about designing and producing artifacts by controlling the structure of matter at the nano-scale. Applications range from cosmetics – nano-sized particles of titanium dioxide are used in sunscreens to give protection without the opaque, white colour of old-style sunblock, to aircraft paint – allowing much thinner coatings of paint to reduce an aircraft's weight. Carbon nanotubes are extremely narrow cylindrical tubes that have a wide range of applications – from strong, light composites that outperform carbon composites, to display screens that will be brighter and sharper – due to the degree of electrical conductivity manifested at this scale. Nanomaterials are likely to permeate our world, but they might not cause the revolution expected. Many nano-enhanced products (such as sunscreens) were introduced without many people noticing their nano-status, and many others have been around for centuries. One area where it may have a real impact is in electronics. Nanotechnology takes electronics to a level of miniaturisation where devices can be 1000 times smaller than conventional semiconductor devices. Electronic nanodevices include new types of very low energy lighting using carbon nanotubes and highly efficient solar cells.

Biomedical engineering: imaging, mending, monitoring

A significant way in which our world has changed over recent centuries is that many of us live comfortably in it for longer. Increase in average life expectancy in the developed world is often cited as evidence of progress and improvement in human knowledge and achievement. Medical science is at the centre of this progress, certainly, but engineering plays an essential and often overlooked role. Part of engineering's contribution follows from the achievements described above: engineering has given many more people a chance to live into old age by improving the conditions in which people live, from removing disease ridden sewage to providing clean water and warmth. However, engineering also has a central role in medicine's mission of preventing and treating previously incurable diseases and life-threatening injuries. Modern medicine is quite simply unimaginable without the input of engineering.

Biomedical engineering takes advances in science and applies them at all stages in medical care – from diagnosis to treatment to long-term health monitoring. Biomedical engineers design, install and maintain products and systems for healthcare applications – from kidney dialysis machines to prosthetic limbs. It is impossible to outline all the ways in which technology enables modern medicine, but some of the major contributions that engineering has made to the key stages of the patient journey can be considered.

The cause of illness, or the extent of injury, is often locked deep inside the body. Finding the cause of disease in a living person was a mystery that could only be solved through inference from stated symptoms and visual examination, or invasive exploratory surgery. The inability to see in detail the causes of illness not only makes diagnosis speculative but treatment inexact – excising a lesion is somewhat difficult if it is hidden from view until the patient is in the operating theatre. However, technologies for imaging the body have revolutionised the diagnostic process. There are very few people who have never had an X-Ray in their lives, as they are used for identifying fractured bones, badly developing teeth and pneumonia shadows in the lungs. X-Rays were famously discovered by Marie Curie, the physicist and chemist, but it is engineering skill that has developed tools for diagnostic X-Ray, making them safer, more exact, more portable; generally continually developing them to meet clinicians' needs. Engineers are also at the centre of more sophisticated (and less damaging) forms of bodily imagining, such as Magnetic Resonance Imaging (MRI). MRI scans clearly show the different tissue types in the body, distinguishing between bone, fat, muscle and pathological tissue such as tumours. MRI is made possible by the understanding and application of magnetism and the quantum mechanical properties of atoms in the water making up a large portion of our body. This application of basic science allows the diagnosis of tumours and, through the development of functional MRI scanners, a better understanding of how our bodies work. Body imaging is not limited to the potentially ill; the use of ultrasound imaging to track the development of the unborn foetus has eliminated a great deal of the mystery and fear that might surround pregnancy and shows how engineering has an input into our health even before we are born.

If any of these myriad devices for imaging the body does reveal illness, surgery techniques for removing pathological tissue have improved significantly through the intervention of engineering. Keyhole surgery allows the removal of tissue with minimal invasion into the patient's body, reducing likelihood of scarring and post-operative complications and accelerating recovery from surgery. Keyhole surgery is carried out with the aid of cameras that are inserted into the patient's body alongside operating instruments. The surgeon works with an image on screen, being able to work inside the patient's body without opening it up. As this area of surgery improves, the surgeon will have access to ever more precise instruments, making moves far more delicate than even the most deft surgeon. One particular improvement is the development of devices that can move in a snake like way, flexibly exploring deeper into the body to carry out surgery or perform investigations for diagnosis. Such devices can potentially rule out the need for external incisions, as they can be used to access the part to be treated through the body's natural cavities.[4]

The improvements in medical care mean that we are living longer and as a result more of us will suffer the illness and inconveniences that beset old age. Medical engineering offers hope for greater support of older people, and a central aspect of this is the development of means for remote monitoring. Utilising communications technologies, clinicians have the potential to monitor people in their own homes, with ever smaller monitoring devices tracking the bodily state and sending crucial information to doctors at remote clinics and hospitals. For example, a device for measuring a patient's blood pressure at regular intervals can be set up to convey readings to a clinician in hospital.

Sensors in a patient's home can track the everyday movements of an individual so that unusual patterns of behaviour that might be due to illness or a fall in the home can be a trigger to alert a doctor or carer. This could potentially change the experience of life for many ill or older patients, allowing them to live a normal life whilst being overseen by their doctors.

The biomedical engineer has to develop an understanding of both the principles of engineering and an understanding of human physiology. This is a fusion of two areas, where engineers become expert in both in order to engineer parts to heal human bodies, or to learn an engineering approach to understanding the body and the ways it can fail. The areas of engineering involved are broad, with important roles for mechanical engineering, electrical engineering and materials science, which all play a crucial role, for example in the design of physical prostheses from a replacement limb (see box) to an

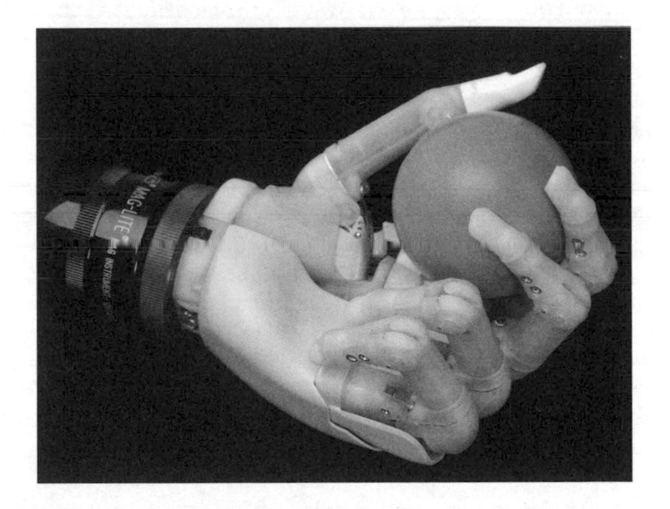

Figure 2 The i-Limb Hand without prosthetic skin covering, showing individually powered digits. *Touch Bionics*

artificial heart. Biomedical engineering is potentially as varied as medicine, however, and biomedical engineers specialise in all of the areas described above, from imaging to robotics for surgery to modelling of biological systems. In all of these areas, biomedical engineers will work closely with clinicians to ensure that their engineering expertise translates into patient care.

THE I-LIMB HAND

The fantasy of a bionic man or woman with replacement body parts that better the real thing is still a somewhat distant dream. The reality is that for quite some time, prosthetic limbs, in particular hands, have been cumbersome and insensitive substitutes for the real thing. The most effective prostheses were often simply claw-like devices which, while helpful to the patient, were by no means a substitute for a hand and all its degrees of movement.

A significant step forward in this area was the i-LIMB Hand, which won the UK's MacRobert Award for engineering in 2008. The i-LIMB Hand was revolutionary in that each digit of the hand was individually powered and able to move independently, giving a significant range of movement and sensitivity of touch. Operated by electrical signals from residual muscles in the arm, the hand is reportedly very easy for a patient to use and comparatively light. It mimics the action of a human hand, articulating around the contours of objects and gripping surfaces with appropriate pressure

The i-LIMB Hand marks a breakthrough in treatment for amputees made possible by the application of mechanical engineering, electronic engineering and the development of materials to make a hand that is closer to the appearance of human skin, and has improved grip by virtue of its skin-like surface. The significance for a recipient is no doubt in large part due to the functionality of the hand, but also the psychological benefit of having a prosthesis that looks and works like a human hand. In this area of biomedical engineering, engineers have the capacity to make a profound impact on peoples' ability to use, and on their perceptions of, their bodies.

3

The mind and methods of an engineer: core elements in engineering

Here, the nature of engineering will be examined by looking at the way that engineers work, the core concepts in engineering and the way that engineers think. It will identify themes that run through all of engineering practice and research; it is the mastery of these elements that enables an engineer to tackle the kinds of complex problems that they have to deal with.

Invention, innovation and manufacturing: from concept to reality

The stereotype of the inventor is a lone individual dreaming up devices that show more ingenuity than practicality. The engineering process, however, is usually quite different from this. As discussed in the biography of Edison, even he, one of the world's most famous inventors, did not succeed through flashes of inspiration experienced alone in a workshop. Rather, his inventions were due to a team of engineers focused on the task of finding solutions to a problem, who were willing to test designs through trial and error.

Invention is a key element in engineering, but by its methods engineering makes a greater contribution to the processes of *innovation*. The Oxford Handbook of Innovation distinguishes invention and innovation thus: 'Invention is the first occurrence of an idea for a new product or process, while innovation is the first attempt to carry it out in practice.' Innovation is all about turning ideas into reality through business or other means which take the early prototype from the drawing board to the user. Innovation is the key to success and riches of individuals, corporations and whole countries, and indeed it was Edison's success as an innovator that gave him his place in history, as his inventions were always developed with a concern for commercial exploitation – he was entirely focused on bringing distributed electricity and electric light to the domestic customer. It is a goal of many governments to encourage and improve innovation within their countries, as it is touted as the key to economic growth.

Innovation is not peculiar to engineering, and any change in the way things are done which successfully makes it through to practice can be understood as a form of innovation. Innovations can be encouraged in retail, in healthcare, in banking. But engineering has a huge contribution to make to innovation. A core activity within engineering is taking the possibilities that are revealed by scientific research and turning them into practical products that can be manufactured, distributed, exported, sold and used. Engineering has at its core the process of turning potentiality into reality. It therefore has innovation at its heart.

There are a number of stages in this process of innovation, all of which require some degree of input from engineering. Research and development (R&D) is about utilising basic research done by the 'pure' sciences in order to carry out applied or directed research towards the end of creating a new product. Development is the process of fine tuning that product into something that really works, by building prototypes, testing and refining. A great deal of engineering research encompasses both sides of the R&D process, focused as it is on producing useful things rather than interesting results to be published in academic journals.

Perhaps engineering's most historically significant contributions have come through developing manufacturing processes so that devices and products can be produced quickly and more efficiently. Without a sufficiently cheap and swift manufacturing process, no invented product can ever make the step from invention to innovation. Engineering has been central to developing manufacturing processes that allow for mass production. Mass production began with the development of processes for making parts of artifacts so that they were interchangeable, and with the development of the lathe for automating the process of manufacturing to increase its speed and to reduce dependence on skilled labour. Marc Brunel (father of Isambard) and Henry Maudslay formed a team which played a significant part in the history of mass manufacturing. A major breakthrough came with a process devised by Brunel for the manufacture of ships' pulley blocks (used in the rigging on ships). Pulley blocks previously needed to be hand-finished by a skilled craftsman, but Brunel devised an automated process for manufacturing the blocks needing minimal skilled work. Brunel's plans were brought to life by Henry Maudslay, who built the machines to put the method into practice. They were met with great success; the factory for making the blocks (in Portsmouth) was able to produce them at ten times the rate of previous methods, each pulley being equal in form and quality. Unfortunately, this manufacturing success was not met with financial good fortune as the customer for the blocks, the British Royal Navy, was not quick to pay up. Brunel suffered similar difficulties with the same client when he devised machines for producing high quality boots for soldiers at a rapid rate.

Brunel's contributions to the manufacturing of boots and

pulley blocks was not a matter of him changing the design of the products in any significant way, but he revolutionised their manufacture in a way that made them far easier and quicker to produce. Henry Ford had a similar impact in the world of motorcars. Ford's fame stems not from a crucial role in the invention of the motorcar but from his huge contribution to making the car available to a wider market. Thus he was central to the automobile revolution, in which the car became the major mode of transport for an increasing proportion of people in the developed world, and the infrastructure needs of motor travel influenced the design of towns and cities. Ford's driving motivation was to make his Model T Ford so easy and quick to manufacture that it would be affordable even to the workers on his production line.

And the production line and its workers were key to Ford's innovations in manufacturing. The Ford approach was to produce only one model of car, and to automate the assembly process so that it could be built quickly without the need for skilled work. Like Brunel he utilised machinery to make standardised, interchangeable parts, and he employed the concepts of the division of labour – breaking down the job of assembling a car so that each worker repeatedly carried out only one small task. This automation of the worker's role was constantly adapted to make it faster by creating a line at which each worker would perform their task so that, instead of working away at their own bench, they would work in a line where the car part to be assembled would move along at fixed intervals from one worker to the next, giving each just enough time to fulfil their task. Thus the worker was part of a machine-like system assembling cars without skill but with accuracy and efficiency.

Ford rewarded his workers with a share of the riches that the Model T brought him by offering wages that were higher than the average. This was part of his vision of producing a car within the reach of the masses; the handsome pay allowed him to create a market for his product. Such means were also essential for him to find the workers to play the roles in his system – the strictly reduced assembly role was monotonous and relentless and no doubt near mental torture for some workers. Yet, the demand for the Ford and the system's ability to meet it meant that the very same model remained on the market for 19 years.

The assembly-line is not the only way that manufacturing was brought into the modern world. Japan and Sweden are two countries that experimented with different models, allowing workers to change tasks, or having small groups of workers complete a whole finished product. The same process does not work everywhere, just as the same product is not successful everywhere. Innovation, manufacturing and indeed marketing are all closely tied to the needs and mores of a particular society at a particular time.

The future might see a move from mass production to manufacturing on a smaller scale. Rapid prototyping is the process of making one-off or small runs of objects, usually by a process analogous to printing. Given the instructions for building a particular artifact, a 3D printer deposits thin layers of plastic, gradually building up a three-dimensional object. The advantage of this process is that prototypes can be created swiftly, rather than by liaising with manufacturers that have the right equipment to build prototypes by a process of machining.

Initially rapid prototyping could only use plastics, but as it has become possible to 'print' objects using different materials including metals, rapid manufacturing has become a reality. By the same process, the finished object can be made rather than just a plastic prototype. The immediate potential for this lies in the possibility of creating products that have niche or dispersed markets, that would not justify large-scale production, as manufacturing can take place without investing in expensive, specialised production equipment. It produces less waste than machining processes which create the desired form by removal of material, and can also be used to make objects that are impossible to create by machining, for example, a single solid object with a hollow in the centre. Potentially, it could create a trend in manufacturing as revolutionary as mass production, reversing the trend from the cottage industry to huge manufacturing plants, and bringing manufacture back to the small scale, yet with the benefits of standardisation, automation and lack of reliance on specialised skills.

GOOGLE – AN ENGINEERING INNOVATION?

The development of Google as the dominant search engine on the web is an example of innovation that is hard to parallel in its success. Though the figures behind Google, Larry Page and Sergei Brin, might not obviously be engineers, the developments that led to Google's success are similar to those in engineering innovation.

Google works because of its use of applied mathematics to solve a problem created by the growth of the Internet and the World Wide Web – that of finding websites relevant to one's interests. Page and Brin locked onto the idea that search results could be ordered in terms of sites' popularity, so that the most high profile, popular and therefore relevant would rise to the top. Their insight was that the number of links to a webpage from other pages would be the key indicator of popularity, and thus they developed the PageRank algorithm to sort through sites. This operation orders websites and pages on the basis of the number of pages that link to them, the PageRank of those sites that link to them, and whether they are one of only a few or many sites listed on the pages that link to them. The popularity of Google is all the evidence needed to show that the algorithm works.

Once the efficacy of the algorithm was proven, huge computer capacity was needed in order to be able to perform the algorithm across the whole of the web. In order to scale up the initial search engine, Page and Brin amassed a farm of PCs that the Google team

GOOGLE – AN ENGINEERING INNOVATION? (*cont.*)

assembled themselves for the purpose. Google is not only a method of searching; it is also the physical hardware that carries out the search.

Page and Brin are innovators because they found a great idea – a way of finding the most relevant pages on the web – and got it to work at the scale required by the ever growing web. And the innovation can be seen as engineering success because it was about identifying a mathematical result that could solve a real world problem, and building the physical system to allow that solution to be implemented. Comparisons can be drawn between Google and the manufacturing innovations of Brunel and Maudslay – a need was identified, a method for meeting it was devised and designed, and bespoke machines were manufactured to meet that need.

Science, mathematics and computing: modelling, predicting, testing

A degree in any engineering discipline will be formed to a large extent by the study of the scientific background to the specialism, and the mathematics required to practice in that area. As engineering has developed, the understanding of how engineered structures and machines behave has deepened, and every engineer has a duty to appreciate the knowledge which has been hard won from the successes and failures of generations of engineers. Accepted engineering science and methods of mathematical analysis are therefore the starting point for any new engineering work, providing guidance in creating a product or system that is safe and functional.

Mathematics has become one of the fundamental modes of representation used by engineers. Take the mechanical or aeronautical engineer working in fluid dynamics. The Navier-Stokes equations are a powerful set of equations in the area of computational fluid dynamics. These equations can describe a number of fluid dynamical processes such as an aircraft or car's movement through the air. The application of these equations allows the engineer to identify how an aircraft of a certain shape will behave in a variety of conditions. Such sets of equations allow engineers to describe the relationships between properties of interest, providing a general, abstract means for describing how systems and structures will behave. Mathematical descriptions of engineering systems are less like ordinary language descriptions and closer to accurately drawn plans. Engineers talk about *mathematical modelling* of a system and the representation of an engineered system in terms of mathematical relationships is akin to building a physical model of that system. Instead of representing its features in 3D space however, the mathematical model represents the system in the space of possibility, and allows the exploration of the system in order to investigate how it will behave.

Mathematics is not merely a means of description in engineering, but an essential tool in predicting the behaviour of the system. Using the Navier-Stokes equations to predict how a particular wing profile will perform is a far more efficient and less costly exercise than having to build and test a prototype for each option under consideration. In structural engineering, an understanding of the properties of materials, such as their resistance to stress and strain, allows mathematical analyses of materials in different structures in order to be sure they can stand up to the forces the structure will experience. These mathematical tools provide a method for running through various design options, assessing which will function best. A good example of this is finite element analysis (FEA).

The metal body of an aircraft can develop small cracks as a result of fatigue caused by the continual compression and decompression it experiences in flight. If those cracks were to grow suddenly, they could cause a catastrophic failure that could end in a crash. Engineers need to know where there are weaknesses and how to adapt the design of the aircraft body in order to prevent damage. FEA is a mathematical method that can give the engineer an idea of how a structure will deform under load. The method works by constructing a grid within the structure being examined and by deriving the likely behavior of the whole structure from the properties and behavior of each element in the grid. Comparing the residual stresses from one element to the next gives an engineer a good idea of where cracks are likely to form. If the engineer can predict where there will be stress concentrations in the structure of an aircraft more time can be spent on improving the design in those weak areas.

Mathematics is important in engineering not only in *predicting* whether a structure will be sound or a machine will work, but also in *demonstrating* that this will be the case. Engineering practised on the basis of pure experience might in many cases be quite successful. And some of the mathematics used by engineers may to a large extent simply provide a mathematical description of what engineers already do. But engineering is done for others and engineered systems will have operators and users. Ability to demonstrate the functionality and safety of an engineered system is essential if the users are to trust in what is delivered. The ability to codify experience in mathematical relationships and to use these to demonstrate the validity of a design is essential in engineering. The application of mathematics in engineering shows that engineering is more than just refined common sense. Sometimes systems can behave in unexpected ways, and having a rigorous means for modelling behaviour can show just why they behave as they do.

Mathematics has the status of a tool in engineering, a method that engineers use to support their work in design and testing. Mathematical relations might be fascinating and beautiful in themselves to the mathematician, but to the engineer they are a means to an end. The mathematics that engineers do might sometimes be somewhat messy and complex, modelling as they do real world phenomena, rather than representing some idealised and purely abstract mathematical relationship. For example, the Navier-Stokes equations are difficult to solve when modelling turbulence, which is obviously a phenomenon of particular interest to the engineer. It is as important for engineers to know where the mathematics breaks down or applies only approximately as it is for them to know where it applies at all.

It is many decades since an engineer has had to rely on a slide rule and a sharp pencil to model the behaviour of an engineered system. Computer-based simulation makes it possible for engineers to run a wider range of simulations, to model more complicated situations, and generally to perform calculations that would be too difficult or laborious for an individual to do themselves. For example, computational fluid dynamics is made far easier by computer programs which perform calculations involving many parameters that would represent days' or weeks' work for one engineer.[1]

Computer models can also take the place of a great deal of physical experimentation, reducing time and cost in the design process. Computer-aided design (CAD) is now an essential method for the engineering designer, allowing the creation of 3D models that can be analysed on screen and altered easily and swiftly. CAD packages distil the expertise of design engineers, allowing the engineer to construct plans using tried and tested solutions, rather than designing from scratch. The development of CAD has made it possible to design structures of far greater complexity, as once a design is created using CAD, it can then be tested by running simulations which will analyse its performance in a wide range of different circumstances.

Software tools have extended the skills and thinking power of engineers hugely, and have no doubt contributed to improvements in safety due to the greater range and detail of the tests that they facilitate. Some worry that reliance on computers for calculation and design brings the threat that engineers will lose touch with experience, and will not develop the ability to simply know when something looks right or wrong. Computer-based calculations might also be seen as lacking the demonstrable rigour of overt mathematical proof, thereby lacking status as a demonstration of the safety or functionality of an engineered system. But, as with mathematics, software is only a tool. The skill of a good engineer is to know the right software to use and the right simulation to run. A good engineer can tell when a slick looking CAD drawing is not quite right. The engineer knows the approximations and assumptions behind the results of CAD and will know when these should be open to question. There will always be a role for experience.

Engineering design: from need to solution

> Engineering design is the systematic, intelligent generation and evaluation of specifications for artifacts whose form and function achieve stated objectives and satisfy specified constraints.[2]

Design is, according to many definitions, the essence of engineering. Design might also be one aspect of engineering that cannot be reduced to mathematical formalism. This is because design is a creative process; as the quote above implies, it is about generating ideas and plans for new artifacts and systems. Engineering design is focused on the existing problems it seeks to solve, but it is free in that there are no pre-ordained, uniquely correct, or even objectively best designs to fulfil a given need. Engineering design was, for many years, absent from engineering education, and many engineers have argued that this was because it is so very hard to teach. While a lecturer can stand at the board and run through the application of the Navier-Stokes equations, how can creativity be taught? How can a teacher describe to a student the process by which they can come up with a novel, interesting and successful design? To a large extent, this cannot be done. Engineering design has to be learned by doing rather than theorising. However, this should not suggest that engineering design is a completely free and irrational process. There are many constraints on the practice of engineering design, and the processes of design in all areas of engineering share many key features.

Engineering design is not a process of free expression, but is highly constrained by various external factors. It has a fixed end goal, which will serve as the test of the design solution. It is in setting out this goal that the difficult problems begin for the engineering designer. If an engineer is designing a product for a client, be that a door for a new model of aircraft or a tiered seating system in a stadium, the engineer has to elicit the requirements from the client. This involves setting out what the design particularly needs to achieve (in the case of the aircraft door it might need to be impossible to open accidentally so that it is safe during flight, but easy to open in the case of emergency), how much the final product should cost, various constraints of size, weight, the conditions in which it will be used (which dictate what materials are suitable) and so on.

The process of setting the requirements for a design project can be fraught with difficulty, and leave much room for error. A client who does not understand the engineering process cannot always identify the most important requirements or those that are actually achievable in engineering terms. The engineer might have a good idea of what would solve the client's problem, but actually fail to have a proper grasp of the real problem at hand. The semi-telepathic process of getting from a need that is in the head of a client, which may be formed in vague, ordinary language terms, into the mind of an engineer who must translate those needs into the language of mathematics and materials science, is open to many mistakes. There are engineers dedicated to just this role for large projects. Requirements elucidation is the skill of correctly identifying the real needs of a client in a way that translates easily into technical solutions. It involves ensuring that what the client wants is actually deliverable – ie feasible (when it comes to software in particular people often ask for things that are not in fact technically possible, overestimating the current state of technology and what it can deliver); possible within the constraints of time and budget; and perhaps most importantly, questioning whether the client is really asking for the right thing – often a need could be met by another solution that is cheaper and easier, perhaps one that already exists. This is an iterative process of feeding potential solutions back to the client, for them to consider against their requirements, and adapting those solutions in light of feedback.

Identifying the set of requirements for a design project goes alongside the conceptual design phase, when ideas of general ways to solve the problems are considered – the more novel the item to be designed, the lengthier and freer this process will be. Once the requirements are set and the concept proven, a more detailed design phase follows which involves specifying a particular form and structure for the artifact and the materials from which it will be made. The result is a set of specifications that set out the features that the final solution must have – what it must do, what it will cost, parameters such as weight, materials used and so on.

After the specifications come the feasibility studies, which involve various kinds of test of the design. This might be in terms of analytic tests, using mathematical models as described above, or computer models as described above; it might be in terms of experiments on a physical model; or it might develop into the construction of a prototype which is tested with respect to the initial requirements. In the majority of cases of engineering design this is an iterative process, with design followed by tests, which suggest refinements to the design, which then demand more tests. The design concept might need to be redeveloped if the idea is shown not to be a successful solution to the initial problem; or the detailed design might be refined if there are significant flaws in the way the prototype performs. Different engineered products will develop differently at this point in the design process. While some involve the design of prototypes that will be tested, others, such as ships or buildings, are complex, large and often one-off artifacts that cannot be tested by prototype. They have simply to be built, once all reasonable testing is done. For those items that are one of many, the design process then involves the design of the production method and facilities needed to actually make the final product.

However, engineering design never really comes to a complete end. Even a finished product can be seen as an implementation of a design that is undergoing testing – 'snags' will arise with any product and the engineer will seek to improve the design. This is obvious from the technologies that surround us: it is not as if someone designed a car – or even a component of a car – and we said 'yes, that works, that's great, next problem'. It is in a continuous cycle of design and improvement, on the basis of how well an artifact functions in use, how easy the manufacturing process was, whether it was possible to maintain and fix it. The laboratory of the engineer extends into the real world. However many tests are performed on computer models or prototypes, it is functioning in real situations over time that is the test of engineering design, and it is in this setting that new design problems become most apparent.

The design process is also open-ended in that there are always numerous ways of meeting a set of specifications, and the designer has to entertain a wide range of possibilities in order to hope to find a good solution. And that is not to say the *best* solution, as there is never one uniquely best solution to a design problem. Competing considerations mean that satisfying one need might infringe on another – for example, making the door of an aircraft more difficult to open in flight might involve adding features which make it heavier.

It is for this reason, perhaps, that for most engineered artifacts it is impossible to come up with the name of an individual who is uncontroversially *the* designer of that artifact. Design is an historical process – almost all design has antecedents and is influenced by successes and failures in objects that are already available. A completely new design of a completely novel product is rare. And even on individual projects it is rarely possible for one engineer to carry out the task alone. Engineering design is multidisciplinary – the aircraft door design will need to take into consideration issues understood best by aeronautical engineers (for aerodynamics); materials engineers (to identify the materials which will be sturdy, lasting and light enough); and electronics engineers (as the door will no doubt be controlled by an automatic control system), amongst others. Each of these teams will involve perhaps tens of people at the conceptual design phase, perhaps hundreds in the testing of specifications. If there is one person leading this, then it is the design manager, but there are likely to be several design managers overseeing different aspects of the work on a complex engineering artifact. Engineering design is therefore a social process, essentially involving teamwork, communication and an understanding of the user whose problem the engineer sets out to solve.

JAMES DYSON AND THE ENGINEERING OF CONVENIENCE

The engineering author James L. Adams wrote this of inventors: 'Successful inventors that I know are extremely problem-sensitive. They are tuned in to the little inconveniences or hardships in life that can be addressed by the technology they know.'[3] This seems a good description of Sir James Dyson, the art school student turned-engineering designer famous for the dual cyclone cleaner (amongst other inventions). Dyson says of himself: 'My own success has been in observing objects in daily use, which, it was always assumed, could not be improved.'[4]

A certain kind of engineering designer is not concerned with a revolution – inventing a completely new type of artifact to meet a need previously impossible to meet. Rather, their concern is the flaws in existing solutions which introduce inconveniences in life that, with a bit of work, could be significantly improved upon.

Dyson's first major design was for an improvement on the wheelbarrow, an unwieldy contraption that does not travel well over rough ground (the kind of ground on which its often used). Dyson's 'ballbarrow' introduced the innovation of using a ball instead of the wheel. The ball gives a larger surface area for the barrow to balance on, and it can tilt from one side to another without tipping. Dyson's ballbarrow performed the same function as a wheelbarrow, looked roughly the same (save, of course, for a lurid orange ball), but, by changing the principle functioning part, did the same job much better.

Dyson's dual-cyclone cleaner is a very similar story. It is intended to do the job of a vacuum cleaner, it looks the part, but, again by changing the principle behind the central working elements performs better. Dyson's epic process of design, product development and testing was set off by a frustration with vacuum cleaners that lost suction as soon as they were used. Vacuum cleaner bags have tiny pores that allow air out while keeping dust in. It is essential that the air is let out, as it is this constant exit of air created by the fan in the cleaner which forces the constant stream of air into the cleaner – the vacuum. But even a little dust can clog the pores and drag the whole process down. Dyson's inspiration for an alternative came from the cyclone that sucks dust out of a sawmill – the

JAMES DYSON AND THE ENGINEERING OF CONVENIENCE (*cont.*)

cone in the roof of a mill that sucks up dust and air and, by the action of centrifugal force, traps dust at its sides and filters it from the air. Dyson's aim was to shrink this process down to fit into a household cleaner – one that would filter dust from the air like a vacuum cleaner but not lose suction in the process.

Dyson's challenge was to develop a cyclone that would handle all of the different kinds of dust and rubbish that a home generates. Mathematical models of the action of a cyclone were available, but these described only the behaviour of particles of a uniform size, rather than the motley muck in the home. Since no mathematical model could handle the complexity of real dirt, experience was the means by which Dyson developed his cleaner – by way of tests on 5,127 prototypes, each changing one feature at a time. The final design enclosed one cyclone inside another of a different shape – the larger outer cyclone collecting string and sealing wax, the inner cyclone collecting fine, well-behaved dust.

There are a few things that are interesting about Dyson's design journey. The first is the starting point – the dissatisfaction with current solutions to everyday problems, and the feeling that they could be addressed much better. The second is the essential role of experience and physical prototyping and testing. Mathematical modelling has been of great importance in performing virtual tests of engineering designs of great scale and grandeur – from bridges to skyscrapers – but for some, even apparently mundane objects, the reality is too complex for the mathematics to describe in a straightforward way. The dual cyclone certainly works on principles well-understood within physics, but these physical laws could not simply be called up to derive a means to apply them in a home cleaner. Finally, although no mathematical analysis could be applied, the process of test and refinement was nonetheless methodical – one change was made at a time, to carefully track those that worked and those that did not.

Dyson's story shows that engineering designs might have their basis in scientific principles, but it takes creative thought and relentless learning from experience to create a design that satisfies everyday niggles.

Systems and complexity: emergence, unpredictability and the human factor

Engineers are increasingly involved in the building of *systems*. That is to say, they are not focused on designing and making one artifact, such as a bicycle, or one part of an artifact in isolation, such as a gear mechanism. Rather, engineers are involved in the design of complex systems with many parts, each of which has to be developed by specialists, but with an eye on how they fit into the whole.

A system is any whole with separate parts that interact and interconnect and affect each other. In an engineered system, these parts will often function in quite different ways and fall into the compasses of quite different engineering disciplines. So, for example, an aircraft is a system comprising many elements, such as the fuselage and wings, the engines, the air compression systems, and the embedded IT systems that control the aircraft. Each of these systems is designed by separate groups of engineers and each involves highly sophisticated understanding. But each must interrelate. The information systems that form the control system in an aircraft have to communicate with the mechanical systems that control landing gear, with the engines, with the door locks and so on. It is rare that any one engineer will have a clear understanding of the detail of how each of these elements work. But systems engineering is the branch of engineering devoted to the design and management of such complex systems, which involve multiple elements, multiple design, construction and maintenance teams, and multiple opportunities for failures and mistakes. Systems engineers have the skill to coordinate the many parts of such complex, engineered objects.

When engineers talk of systems, they also talk about complexity; this denotes a peculiar feature of engineered systems that poses particular challenges. A complex system is not just a complicated system – a system made of very many parts that are related in ways that it might be hard to recall and to track. The parts of a complex system are related in such a way that the behaviour of the complete system is greater than the sum of its parts. Their interaction gives rise to features and properties that cannot be accounted for by the properties or behaviour of individual parts. If something is merely complicated, a model aircraft say, it can still be taken apart and put together, and the whole finished product is defined entirely in terms of the many tiny parts and how they are assembled. A complex system, however, cannot be analysed into its parts in this way. It will have what are known as *emergent* properties, which result from the way that those parts are assembled. A complex system is, if you like, more like a cake than a salad. Putting all the ingredients together does not just give you a combination of all those ingredients, it gives you a product which has properties that the original ingredients did not.

The interactions between the different parts of a complex system account for these emergent properties. This poses a challenge because it is not always known exactly how parts will interact and what the effect of those interactions might be. Systems engineering is all about understanding the nature of systems in order to design for desirable emergent properties and to exclude as far as possible interactions which could cause the system to fail.

Systems engineering is the principled practice of designing and maintaining complex systems. Its roots lie in the defence sector, which both employs highly complex technologies such as fighter aircraft which exhibit all of the features of systems described above, and which employs such technologies in complex operations involving many of these technological artifacts in conjunction in carefully planned operations involving many players. Defence involves the use of complexes of technology: missiles carried by fighter jets, fighter jets carried by aircraft carriers. A warship is not simply an enlarged dinghy, it is a system that performs many functions from performing as a mode of sea transport to carrying other vehicles and munitions and serving as a platform for military operations. Therefore, the ability to think in terms of and to design systems with interrelated and technologically diverse parts is integral to modern defence.

Systems engineering is not limited to the defence sector and as technology develops the need for systems thinking is required in many more areas. Transport networks are complex systems. They are composed of physical infrastructure, such as the road networks; information infrastructure, increasingly to monitor the traffic levels on roads and to send that information to road side displays; and the motor vehicles themselves which merge mechanical engineering with software engineering through the engine management system. We also talk of computer or software systems engineering, and systems thinking has become increasingly crucial in software engineering. As information systems come to support more and more of our daily activities, from the transport networks to national databases of patient information, software has to be considered as part of a system of disparate technologies, and as a system in its own right supporting several disparate (and even conflicting) functions from collating and comparing data to ensuring its security.

These examples should show that a central characteristic of systems engineering is that it is interdisciplinary. The systems engineer cannot define him or herself as a mechanical or civil engineer, as systems engineering involves being able to understand and appreciate the challenges of engineering tasks in all engineering disciplines. Although systems engineers cannot be expert in all things from designing engines to IT systems, they will have to know a sufficient amount about each to understand how they will interact and to be able to build bridges between the work of the engine designer and the software engineer. The systems engineer is an engineering polyglot, able to understand

and translate between engineering specialisations, and able to express the needs of all of them in a common tongue. The systems engineer does not replace the specialist, although in many cases will, or should, be expert in at least one aspect of the system in question, but coordinates the activities of those specialists. Sunny Auyang calls systems engineering 'the design of a design', as what the systems engineer does is to plan how all of these disciplines will work together designing parts of a system in such a way that the whole emerges as a coherent system which performs the functions intended of it.

The systems engineer has, therefore, to be able to maintain a holistic attitude towards the technologies he or she develops. They have to think of the whole, rather than thinking of one part then another. But they also have to think of a system in terms of its whole lifecycle, from cradle-to-grave. The complexity of an engineered system means that the interactions between its parts create a whole that cannot necessarily be taken apart in just the way it was put together. You can take a salad apart to remove the celery if you find your dinner guest does not like it, but you can't remove the egg from a cake. Similarly, the parts of a system as described in its design phase may be put together in such a way in the process of construction that it is not easy to take them out if they fail to function or if they need maintenance. This also means that changes in the design of a complex system are not straightforward. You cannot change one element without affecting the whole. So systems engineering involves understanding and managing the development of a system through design and construction to use and maintenance. And the grave stage of a system is of great importance. A system has to be taken apart safely at the end of its useful life, in such a way that its parts can at best be reused or recycled and at the very least disposed of without causing pollution. All of these stages in the lifecycle of an engineered system must be envisaged and planned for at the initial design phase.

A crucial aspect of many systems which is implicit in the discussions above is that they are *sociotechnical* systems. That is to say, they incorporate both bits of physically engineered technology, and the people that operate, maintain, use or otherwise stand in the near vicinity of those systems. The transport network described above does not only involve roads, cars, GPS systems and roadside alerting systems; it crucially involves drivers and passengers. Aircraft are a complex mesh of not only engines, wings, fuselages, embedded computer control systems and other elements, but also passengers, crew, pilots, air traffic control staff and the ground crew at the airport. These are not passive elements, but integral active parts of the system that have a huge influence over how it will function. How drivers will react to congestion on the road determines how the congestion is managed; how they will react to a message on their dashboard determines how faults in a motor vehicle will affect its performance over time. How passengers will behave in the instance of a serious failure or crash in an aircraft will determine how well they survive that failure. So, human behaviour is central to the function and safety of engineered systems.

The design, construction and delivery of an engineered system is also a social enterprise that is hugely affected by the behaviour of and interactions between the teams of people involved in each part of the system. The Airbus A380, for example, the largest passenger aircraft at its launch in 2007, was the result of a collaboration of teams of engineers across Europe, with wings made in Wales, fuselage in Germany and final assembly in France. The coordination of a team of teams divided by geography, specialisms and language is a challenge quite different to, but inextricable from, the demands of physically engineering the parts of the aircraft.

Uniting the technical and human parts of systems is a great challenge for engineers. It involves quite different kinds of understanding, as the two elements are governed by quite different laws

– the physical laws that govern aerodynamics and the psychological laws, if there are such, that determine the behaviour of people. Philosophers have worried for centuries about how the mind, which is invisible and intangible, interacts with the physical body which is seemingly governed by the laws of physics. Sociotechnical systems form a large-scale analogue of this problem, one which is, unlike the mind-body problem, in pressing need of resolution. Human actors introduce risk into systems through either careless accidents (losing attention at the wheel), or natural behaviour (the urges and instincts that kick in when evacuating an aircraft). Engineers need to understand, or work with people who understand, those behaviours in order to design sociotechnical systems effectively and safely.

Systems do not exist only in the technical realm and systems thinking is not confined to engineering. As mentioned in the last chapter, systems thinking is entering into biology, with the human body considered as a complex of interacting parts rather than a concatenation of cells. The engineer that develops system skills will be in high demand, as such skills are invaluable in managing all kinds of complex systems, from hospitals to financial institutions. Systems thinking is an attitude that is invaluable in an increasingly complex, networked and global world.

Risk, uncertainty and failure: controlling the future

'Structural engineering is the art of modelling materials we do not wholly understand, into shapes we cannot precisely analyse so as to withstand forces we cannot properly assess, in such a way that the public has no reason to suspect the extent of our ignorance.' This often quoted characterisation of engineering is due to Dr A R Dykes, Chairman of the Scottish branch of the UK Institution of Structural Engineers, whom he was addressing in 1978. Although it appears to satirise engineering method, it very nicely brings to the fore a key aspect of engineering practise – the unavoidability of risk and uncertainty. Dykes was talking specifically about structural engineering, in the context of the risk of computational techniques for analysing the likely behaviour of structures. His point was that structural engineering was still an art rather than a science, not reducible to completely predictable and controllable principles. What he says applies just as well to other areas of engineering. And it makes a point about a key aspect of engineering – the fact that engineers frequently work with a significant element of uncertainty.

Engineering practice is not like scientific research, carried out in the lab under carefully controlled conditions that can be varied by the experimenter. Engineering takes place in the real world, where numerous influences interact and where circumstances are frequently novel and variable. As a result, the engineer cannot always predict with certainty how an engineered system will behave in all of the situations it will face. This uncertainty means that engineered systems always contain an element of risk – risk that they won't function, risk that they will fail dramatically, risk that the costs of a system will escalate, and the risk that they will turn out to be a huge waste of money. Therefore, key to engineering is the ability to assess and to manage the risks inherent in any engineered artifact, process or system.

What do engineers mean by 'risk'? There are a number of related concepts that need to be teased apart to understand what engineers think about when they think about risk. *Risk* should be distinguished from *hazard*. A hazard is the kind of danger posed by a substance or process, for example a hazard associated with a nuclear power station is that it the reactor core could overheat and produce a fire or explosion. This is a hazard that we have seen come to pass before. However, the risk associated with the operation of a nuclear power station is the *likelihood* of that particular hazard coming to pass. Therefore, although nuclear power stations might be seen by some as particularly hazardous places because of the kinds of accidents that can occur they pose in fact quite a low risk as the likelihood of such an accident is quite low. This can be compared to situations which pose a much less severe hazard but the likelihood of that hazard occurring is very high, for example, the likelihood of being cut or burnt if you work in a busy kitchen. But it does not follow that because the chances are high in this latter case that kitchens are objectively riskier places than nuclear power stations. In addition to the likelihood of a given hazard, a crucial factor in assessing risk is the impact of that hazard. The potential impact of an explosion in a nuclear power station is very high – it can cause the contamination of a wide area of land, can kill people involved in the accident and cause serious health problems for those in the environs. Therefore, it makes sense to think of a nuclear power station as a riskier place to be because although the likelihood of accident is low, the impact of such an accident is far greater than the highly likely but utterly tolerable risk of cutting oneself with a knife in a kitchen.

Engineering risks often arise in situations about which there is a great deal of uncertainty. Some of the greatest hazards are posed by situations that happen very rarely, such as an explosion in a nuclear power plant or the collapse of a building. Although it might be possible to calculate the likelihood of these things happening on the basis of an understanding of the likelihood of each event that would lead up to the outcome, the complexity of such situations (see previous section) makes any such calculation questionable. Such low likelihood events do not form the basis for statistical calculations of risk as they are, fortunately, so low in occurrence that determining the probability of them occurring is impossible. There are also circumstances in which there is uncertainty about the nature and impact of any potential hazard. For example, if there is an accident in a nuclear power station which causes a leak of radiation, it may not be possible to predict in detail the effects of the radiation, how far they will spread and how long they will last. Engineers have to cope with these levels of uncertainty, either by seeking to reduce the uncertainty by understanding the processes that would lead up to an accident, or by taking into account this uncertainty when acting on a calculated risk.

Engineering risks do not concern only the risk of accident, but also financial and business risk. In an engineering project, there is always risk of failure, risk of exceeding a budget, risk of not being able to complete a project on time. These forms of risks also have to be brought under control by an engineer, as a great deal of harm can be caused by taking on a project that it is almost impossible to complete. For example, a risk is taken if an organisation invests heavily in an IT system. There is a risk that the project might be more complex and costly than first forecast, and the cost in this case is in both lost money and the hazards posed by a poorly functioning system. The engineers and other professionals delivering such systems have a duty to set out risks of this nature.

It is therefore a key part of an engineer's job to identify ways in which systems can fail. Identifying failure modes is the first step in managing engineering risk. A number of strategies are then employed to diminish the risk of failure. In areas such as civil and structural engineering, the amount of weight or stress that a structure can withstand before it collapses must be assessed. It is essential that the structure can withstand far more stress than it is likely to face in its operational life. If a lift, say, is designed to carry ten people of average weight, it might in fact be designed so that it will carry 30 people of average weight before it is likely to suffer failure. This means that it is designed with a factor of safety of 3; it is able to carry 3 times the weight ever expected of it. The ways that an engineering system can fail will also be identified, so that there can be back up processes established that will continue to work even if failure occurs. Engineers should ensure that there are no common failure modes for these processes; that is, the back up should not be vulnerable to the same problems as the main system.

The key thing is that engineered systems should be designed to withstand uncertain future situations. This is what makes engineering more than simple construction or tinkering – engineered systems should not simply be put together to work in a given situation, they should be designed to function in a variety of situations and be robust and reliable. Highly critical engineered systems, such as the power grid, or the telephone network, will be engineered so the likelihood of failure is extremely low, as the impact of failure in such a system would be intolerable.

This is the ideal, of course, and not all engineered systems will be perfect. Failure is a fact of life that engineers face probably more than the rest of us. But failures have silver linings for engineers, as they offer lessons into perhaps undetected weaknesses in a design, and give guidance for future designs. The flight recorder in an aircraft, for example, is a useful engineering tool as it gives clues as to why a crash occurred, and whether it is due to a design or construction fault that could be avoided. A large part of structural engineering is concerned with understanding why structures fail, so that more robust structures can be built. Whole books have been written about engineering disasters and what was or should have been learnt from them (Henry Petroskis's *To Engineer is Human* is an example), and particular disasters such as the failure of the space shuttle *Challenger* and the collapse of a walkway at the Hyatt Regency Hotel in Kansas, 1981, have been pored over in books and documentaries to warn engineers and the wider public of the need to foresee failure and to take decisive action whenever a failure is envisaged.

How far does an engineer have to go to prevent failure and

diminish risk in an engineered system? How does an engineer establish that a risk has been reduced? There will always be trade-offs involved here, as reducing risk will often involve increasing the cost of a system, and sometimes its complexity. This financial risk has to be balanced against safety risk. And reputational risk also comes into the balance – what is the harm done to an engineering company if a risk is taken and there is an accident? The steps that will be taken depend to a large extent on what is considered to be an acceptable level of risk for a system. Unfortunately, establishing what this level is might not be straightforward.

Engineers often become frustrated because their understanding of risk does not tally with the public's perception. For example, it is far less likely that a person will experience a fatal accident if they take a journey by air, than if they take even a short journey by car. Yet people tend to perceive air travel as far riskier than travel by road, even though the probability that they will experience an accident is exceedingly low. This is because there are qualitative aspects to risk which need to be taken into account, especially when controlling risks that will affect the public. People's attitudes to different kinds of risks are not only determined by their likelihood, but by other quite common features. People might fear air travel because they are not in control in the way that they would be if they were driving a car. They may also fear it because accidents in air travel have a more catastrophic outcome – it is often the case that when there is an air crash hundreds of people will die, whilst road crashes often involve no or few deaths. A sense of unpredictability of a threat and its remoteness from one's control is central to people's reactions to risk. It is perhaps for this reason that after the bombings on the London Underground in July 2005, many people took up cycling for their commute. However, the likelihood of experiencing an accident on a bicycle, especially for an inexperienced cyclist, is far higher than the likelihood of being involved in a terrorist attack. Still, many people find it far more acceptable to take a risk that they feel is under their control and which they voluntarily expose themselves to.

Although these attitudes might seem irrational to the engineer who is accustomed to reducing risk to quantitative assessments of the likelihood of an accident, there are understandable reasons behind these attitudes and they form part of common sense rationality. Engineers have to work with these common sense attitudes to risk in a great many applications because the risks they deal with affect the wider public. As a result, dilemmas can arise that are hard to resolve. For example, as with air travel, people are less tolerant of accidents on railways than they are of accidents on the roads. Therefore, there is considerable public pressure to continually work to make railways safer. But there are limits on how far this is possible. It costs money to make safety improvements, and to make a railway safe might therefore mean financing such improvements by huge ticket price hikes. How tolerant are people of this impact of risk reduction? This brings into the equation the value of reducing fatalities or accidents on the railway. Someone must decide how much it is worth spending to reduce the likelihood of a fatal accident. Can we entertain such thoughts without feeling that we are moving toward putting a price on a life? And of course, the reduction of risk to the passenger on a train is going to be the result of further work on the railway system, often meaning the employment of more maintenance workers on the track. The increased safety of the passenger might therefore be accompanied by increased risk for workers.

People's different attitudes to risks that pose a catastrophic outcome and to risks of lots of lesser hazards over a long period of time have to be considered when choosing between the development of nuclear power stations for energy generation over fossil fuel power stations. Acceptability of the former might be influenced by the fear of a catastrophic accident – but the preference for the latter has to be qualified by the contribution such stations make to climate change, raising the risk of lots of very negative effects over time. Is some level of risk essential for progress in engineering? How far can risks be reduced? Daniel Gooch said of I. K. Brunel: 'Great things are not done by those who sit down and count the cost of every thought and act.' Although some risk might be unavoidable, it is now considered completely essential for an engineer to count the cost of every thought and act. Taking risks without taking them seriously is unacceptable. But the hard decision is what to do once the cost is counted. How much risk should be tolerated, and how much should we spend to eliminate it? These kinds of extremely difficult questions will be examined in the next chapter.

4

Social constructions: engineering and society

Chapter two outlined the major disciplines of engineering in terms of the ways that they have shaped our world – from physically altering the landscape, to extending human reach into space, to constructing a virtual world. The changes described in that chapter are obviously not confined to the physical world; engineering has also revolutionised many people's experience of the world and the ways that they live in it. The idea that transport networks and communication technologies have made the world smaller really relates to the way in which it feels like a more compact place because long distances can now be travelled relatively quickly. Engineers have helped to create a virtual world in that they have created the means by which people can experience relationships and carry out daily activities via new media, which allow us to have social relationships built on new roles and new ways of transacting.

However, in that chapter there was no deep exploration of the ways in which engineering has changed our lifestyles and our quality of life. This chapter will explore the ways in which engineering has shaped society, and the influence of society and individuals' choices on engineering. It will look at some of the negative effects of engineering and how it has disrupted our lifestyles. Engineers are becoming ever more keenly aware of the wider implications of their work and great effort has been invested over decades in producing codes of ethics for engineers and for individual engineering institutions. This chapter will end with a discussion of engineers' ethical responsibilities and what they are, or should be, doing to meet those duties.

How engineering shapes society: sewers and Silicon Valley

The daily lives of inhabitants of developed countries are infused with the products of engineering, and there are increasingly few aspects of life that are not influenced by engineered inventions or processes. The very fact that so many people in the western world live in homes with clean water for drinking and linked to sewage systems for removing wastewater, the fact that they are connected to a power grid that means they can switch on the light or the TV whatever time of the day or night, and that they can contact relatives, friends and co-workers by telephone and email is nothing less than an engineering miracle. If one meditates on the changes that modern life has undergone by being connected to water, power and communications infrastructures, it seems that engineering and technology have had as profound an influence on human life as any political revolution or the development of any religion. The very process of connecting homes and businesses to the services that we all now rely on is a phenomenal achievement when one considers that the majority of towns and cities in the developed world were established before the invention of large-scale electricity generation and distribution, before the invention of telecommunications, and even before the development of modern water supply and sewage systems. Retrospectively refitting the city to bring essential services to its inhabitants has reshaped the places we live in, turning streets once connected only by bricks and mortar into homes linked to power, water and communications networks.

To illustrate the impact that engineering has had on life in large cities, consider the contribution of sewerage to improving quality of life. Before the development of its sewage system, London suffered devastating outbreaks of cholera, as its inhabitants drank water contaminated by their own waste. Without an understanding of the means by which cholera was spread into drinking water, the problem that London's inadequate sewer system posed to public health was not properly appreciated. What could not fail to be recognised, however, was the effect that the overflowing cesspools of London had on its atmosphere. The heat of a particularly warm summer in 1858 cooked up an unbearable stench from the sewage that fed into the Thames. The smell creeping in to the Houses of Parliament meant that the politicians who had procrastinated over the problem of London's sewage were forced to act.

Joseph Bazalgette was the engineer behind the sewer system that saved London. As Chief Engineer of the Metropolitan Board of Works, he was leading a project that had stagnated while London's bureaucrats failed to take action. His plans for a sewage system involved a series of major sewers on both banks of the Thames that would be fed by the existing sewers formed by streams and drains leading to the Thames. These major sewers were designed so that they sloped down to the east, gravity pushing sewage out of the city. Far enough from the centre of population, pumping stations would raise the sewage out of the deeper tunnels, and would release it into the Thames at ebb tide, so that nature's own flushing system could carry it to sea. By 1865 the first phase was completed, with the pumping station at Crossness, south east London, brought into action. The second phase was marked by the opening of Victoria embankment in 1870.

London's centre features wide walkways on either side of the river, the Victoria and Albert embankments, which are mere feet above the tunnels that wash Londoners' effluent out of the city. Despite their dirty work, they are attractions of London, forming tree lined promenades along the river. The pumping stations in East London are also things of beauty, highly decorated and in their time brightly coloured, they are testament to the pride in the sewerage project, pride both in its enormity and the importance of what it achieved. Of course, there was much more to be done to improve the living conditions of London's poorest inhabitants, but London's sewers demonstrate engineering's capacity to have an impact on the quality and length of life in large communities.

Over two centuries have passed since London developed its sewer system, and other cities such as Paris completed similarly ambitious projects. Despite this, the transformative effects of modern water supply and sewerage systems have yet to affect many parts of the world. Some of the world's poorest people live in informal settlements that have no infrastructure and therefore no sanitation systems and as a result suffer from the illnesses caused by contact with human waste. The large-scale engineering projects such as those of London and Paris are inappropriate in these settings. However, engineers can help to address their particular problems by finding ways to create cheap, robust and portable technologies for removing waste water.[1]

Not only can engineering transform existing cities and towns, there are centres of population in the developed world that became established only as a result of developments in engineering. Just as ancient civilisations settled where there was abundant water or food, modern towns and cities have grown where industry has flourished. As industry has found abundant use for natural fuels like coal and oil, and devised the means for extracting these resources by intensive methods, so communities have grown and flourished through following the opportunities presented. Mining towns and cities of oil are developed not simply because those resources are there, but because there are the means to create an industry out of those resources through large-scale mining, extraction and processing of natural resources.

The development of transport links not only made connections between existing habitations but also gave reason for new towns to develop. Railway towns developed during the period of 'railway mania' in the UK, places like Crewe and Swindon growing up because they were sites of stations where many trains stopped or where connections were made, providing opportunities for businesses to make the most of passing travellers. In the US, the transcontinental railroad connected remote and previously uncharted territory and towns were built hurriedly in otherwise inhospitable locations on the rumour that the railway would be passing through a given point. Films like *Once Upon a Time in the West* capture the frenzy amongst speculators hoping to make money out of the railway, and of course, the potential for corruption and violence when there is the opportunity to strike it rich. Industry and construction creates its own communities, from gritty Boulder City which housed the workers on the Hoover Dam to the rather more salubrious Silicon Valley in California.

The development of industry does not simply influence where people live, of course, but has meant huge changes in people's way of life since the Industrial Revolution. The process of industrialisation through the mechanisation of manufacturing has no doubt been one of the great shaping influences in modern society. The developments initiated by Brunel and Maudslay's innovations led to a decline in cottage industry, where products were made in small quantities by skilled craftsmen to meet the needs of the immediate or nearby communities. Instead, workers were employed in factories overseeing the work of the machines, acting as semi- or unskilled links in the chain of automated machines that performed the discrete tasks in the manufacturing process. Ford's factory, as described in the last chapter, was a paradigm example of this change. In particular, it was an example of how these large, centralised manufacturing centres attracted workers who could now find jobs many miles from their homes without the need for particular skills or training. The Ford factory was the workplace of many migrant workers, with 70% of the 1914 workforce comprising 22 different nationalities. This paradigm is repeated in the automobile and other industries across the US and Europe.

The march of automation continued throughout the twentieth century with even more of the tasks of manufacturing taken over by machine, with the rise of the assembly line robot, leading to fewer people being employed in the unskilled tasks of the assembly line. Changes in manufacturing continue to affect peoples' working lives. Perhaps most notable now is the move of manufacturing to developing countries such as China with a decline in manufacturing industry in countries like the UK. The loss of an industry can have as profound an effect as its introduction – with communities in the vicinity of closed-down mines and factory plants feeling the loss of the industries that once gave them cohesion and a good wage.

The developments of engineered technologies continue to influence the way people live and work. The development of the increasingly powerful and increasingly cheap personal computer and the infiltration of the Internet through broadband and wireless connectivity mean that people who might usually work in offices potentially have greater choice over where they spend their working day. The demise of manufacturing in many countries has been replaced by the so-called service and knowledge economies, where money is made through providing business services and selling ideas rather than making 'things'. In this context, there is less need for businesses to be located in specific areas or for individuals even to have to work together in the same location. The prediction that we might all one day work in a less structured way and that the commute to work will become obsolete has been made many times. But every time the prediction is made, personal computing power has increased a little more, as does the bandwidth available for connectivity and communications. As the technologies improve for supporting remote conferencing, and as communicating through ICT platforms such as blogs, social networking sites and multiplayer online games becomes the norm, so the rise of the virtual workplace becomes a real possibility. However, whether these possibilities are realised will be determined by a range of factors, as will be discussed shortly.

Making fun

The impact of engineering on our working lives is a complicated matter, with overall improvement overshadowed by tedious and risky working conditions for many, but the contribution of engineering to our social lives is rather more positive. While it might be obvious that technologies like MP3 players are engineered artifacts, engineering permeates our leisure hours in many ways and has done for decades – even centuries. The Renaissance craftspeople, including Leonardo da Vinci, applied their mechanical skills to construct elaborate automata that would be displayed and marvelled at in public celebrations. Public and private leisure hours are now ever more strongly influenced by increasingly sophisticated engineering inventions.

Music can be a pure and uncomplicated pleasure, and is as ancient as birdsong, but the creation of music and our experience of it are increasingly mediated by technology. If you play an acoustic instrument and much of your musical pleasure comes from hours of playing alone or in an orchestra, then for you music might be a refuge from the technology-dominated world.

But for the majority of music lovers, the experience of music is made possible, either directly or indirectly, by the products of engineering.

Recording and playback of music is obviously an engineering innovation. Thomas Edison produced a device called a 'phonograph' which made recordings onto a wax cylinder and performed rather shaky playback. Some of his experimental recordings, including a rendition of 'Mary had a Little Lamb' remain, though they may not convincingly demonstrate that engineering is the source of musical pleasure. But the wax disk and the phonograph that played it was the parent of many generations of new musical media, from the vinyl record and gramophone, through magnetic tape, CD and the flash drive of an MP3 player. These developing formats have enabled better quality and more authentic playback, greater capacity for storing music, and have made it easier to acquire and share recordings (though some aficionados would argue that the greater capacity gained by recent developments have been at the cost of the authenticity and quality of sound afforded by vinyl).

Formats for recording and playback have been essential in the delivery of music to its audience, but the making of music is itself an increasingly technological enterprise. Since the electrification of musical instruments in mainstream music from the fifties onwards, engineering has underpinned the search for better, more radical sound. The early evolution of electronic music was aided by composers for TV and film who were seeking to create otherworldly sounds to accompany increasingly fantastical celluloid visions of outer space. As society became gripped by the concept of space, it demanded a visual and musical representation of what an intergalactic future might look and sound like. Among the early pioneers were Louis and Bebe Barron who created the first entirely electronic film score for the film Forbidden Planet in 1956. From this tradition was born the BBC's groundbreaking Radiophonic Workshop which opened its doors in 1958.

Experimentation with electronic music was pushed further in the 1960s and '70s with the development of the synthesiser. The name might suggest that the intention of the synthesiser was to emulate natural sounds and recreate them by electronic means, but the real impact of the synthesiser was in bringing to music sounds that are completely out of this world. The sound of the synthesiser was like candy to the psychedelic generation, and the synthesiser was essential to any band wanting to create that 'far-out' sound. Some artists that were most intrigued by the scope of the synthesiser started to build machines of their own. 'Tonto's expanding headband' was the soubriquet of Bob Margouleff and Malcolm Cecil, with 'Tonto' referring to a monster machine comprising modules from various synthesisers available at the time. Working as sound engineers they used this construction in the production of albums by Stevie Wonder. Another musician to explore the world of the synthesiser was jazz keyboardist Herbie Hancock, who perhaps had an affinity for it because he started out on a college education in electrical engineering. Through the 1970s and '80s bands like Kraftwerk, for whom the line between building equipment and making music was blurred, continued to explore the world of music created by electronic means.

Modern music composition and production owes a great deal to software development as well as to electronics, with readily available software for sampling and sequencing facilitating the production of home-made music by amateur musicians. As information technology has democratised many things, it has democratised music, with recording artists able to make their own music, upload it onto the web and share it with a world-wide audience.

BOB MOOG – THE ENGINEER PEOPLE WEAR ON THEIR SHIRT

Robert Moog (1934–2005) started out tinkering with radio equipment that his father, an engineer, kept in a basement workshop. Moog got interested in building theremins, an electronic instrument invented by Louis Theremin which was popular in the creation of music for science fiction films. His first business venture was creating kits for self-asembly theremins. From here he became interested in new ways to make electronic music.

The challenge which motivated Moog to develop the synthesiser was that of creating electronic sounds with varying pitch. The sounds in a synthesiser are generated by an oscillator, and Moog was able to move the sound created by the oscillator through its pitch range by controlling the voltage being fed into it. Another key feature of his synthesiser, was the 'envelope generator', which could create a controlled burst of sound so that a note could be struck by pressing a button or key.

The Moog synthesiser was a collection of separate modules incorporating the voltage-controlled oscillators, the envelope generator, along with filters and the amplifier, which could be wired together in different ways to create different sounds. The flexible connections were created by 'patch cords', and different 'patches' between the modules created different sounds. The modules in the Moog synthesiser were thus manually connected and re-connected by the musician playing it. This meant that making music was an engineering task in itself. Very often the engineers who helped to build Moogs had to teach musicians to play them, or even play the synthesiser for them.

The Moog was popular with artists like Stevie Wonder and Emerson, Lake and Palmer, but the recording that really pushed the Moog into the public eye was a set of recordings of Bach, entitled *Switched-On Bach*, by Wendy Carlos. It seems that public acceptance might have depended on proving that the Moog was capable of playing 'real' music.

Bob Moog succeeded in merging his engineer's desire to tinker with the musician's desire to create innovative music. The sound of the Moog has become iconic and the Moog logo, familiar from the back of the synthesiser, is now commonly seen on T-shirts. This is certainly a rare accolade for an engineer.[2]

A similar story can be told about the development of the performing arts from stage plays to film and television. From tape reels to video to DVD and now high definition DVD (HD DVD), and from projection to cathode-ray tubes to flat-screen high definition televisions, the ways that we experience storytelling has undergone transformations due to engineering breakthroughs. From the sound, lighting and staging systems in theatres, to the relatively cheap formats and platforms which allow people to watch films and plays in their own home, engineering has a significant role in making entertainment accessible to an ever wider audience.

Now of course, from the invention of the television set to the pervasion of digital broadcasting, engineering effort is focused more directly on bringing us entertainment. This is because broadcasting has become part of our essential infrastructure – in countries with free media, people rely on television and radio for news and for information, especially in times of crisis or historic celebration. Some of the most significant events in human history, from the first steps on the moon to the terrible events of 11 September 2001, united people worldwide who listened to or watched events unfurl from their own living rooms and offices. Mobile communications can also have a profound role in sharing information. The pervasion of mobile phones in Africa is sometimes credited with making it far harder for corruption to go unreported, with voters able to rapidly communicate election results to radio stations making it harder for authorities to misreport them. As mobile multimedia grows, with handheld devices connecting us to the Internet and digital broadcasting wherever we are, we need never be disconnected from the information and entertainment infrastructure.

These examples show how electronics and information and communications technologies are essential to many people's leisure hours, but more public forms of fun are dependent on other areas of engineering. Structural engineers bring us the sports stadiums that host live sport and stadium rock. The Millennium Dome in London, for example, was built to mark the dawn of the twenty-first century, and was hailed as a great engineering achievement with its extremely light, tent-like roof stretched over a dome-shaped network of cables. Unfortunately, the millennium exhibition that it housed turned out not to be so entertaining and the expensive project threatened to be a major political embarrassment. However, the engineering firm, Buro Happold, who performed the innovative work on the dome's roof were drafted in again to help develop it into a concert venue. In its new guise as the O_2 arena, it is one of the most successful venues for music in the world, in 2008 overtaking ticket sales at Madison Square Gardens for live performances.

The kinds of structural innovations involved in the design and construction of the Millennium Dome are often required by the art world to construct groundbreaking public art. Visitors to the north of England are heralded by a vast sculpture, the *Angel of the Nort*h, a human form with outstretched arms represented by aircraft wings. Standing at 20 metres and with a wing span of 53 metres, Antony Gormley's sculpture demanded the skills of structural engineers to give it a firm foundation and to ensure that the artist's vision remained true at this scale. And of course, at this size, erecting the sculpture was a significant construction project. The engineering firm Arup were able to meet these challenges, and have brought other transcendent artistic visions to reality, including a blood red, 150 metre wide, stretched PVC membrane named *Marsyas*, created by Anish Kapoor for the Tate Modern in London.

Engineering expertise is also essential to structures with a rather more visceral impact. Rollercoaster design employs the skills of a range of engineering disciplines. A rollercoaster designer will set out by establishing the track geometry, the route that the ride will take, followed by a dynamic analysis of that route. A rollercoaster car has no power and no steering. An

Figure 3 Anish Kapoor's sculpture *Marsyas*, exhibited in the Tate Modern, London in 2002 and constructed with the assistance of engineers from Arup. *Photo © Arup*

electric motor raises it to the top of the 'lift hill' at the start of the ride, and after that it makes its stomach-flipping journey under the effects of gravity alone. The dynamic analysis maps the speed and accelerations of a vehicle following the route, and this establishes whether it is the right length and shape to take the rollercoaster car all the way around without running out of momentum. A structural engineer designs the track and its supporting structure, ensuring that it can withstand the forces of the elements and the car that rides it. A mechanical engineer works on the car, designing it for fatigue loading, so that it can withstand the stress of running at speed on hard wheels without the protection of suspension. The car has to be engineered not to jump or tip off the track and it has to contain its passengers so that they are not spontaneously ejected under the force of gravity. Designing the restraints in the car so that they hold each passenger in place snugly and comfortably is a challenge when passengers range from slim teenagers to body-builders.

Rollercoasters have to be designed for thrills but also for safety. If a ride has more than one car, so that one is loaded while another travels around the track, there is always a chance of collision. Electronic control systems monitor the journeys of the cars and link to safety systems that step in if anything is out of the ordinary. The design of these systems is a complex matter. The designers of the rides have to consider in advance all the ways that the ride could fail so that control engineers can programme all of these scenarios into the system – this includes incorporating back-up controls in case it is the control system itself that fails. Multidisciplinary teams are also involved in thorough testing of the rollercoaster once it is in place, running through all possible failure modes to make sure that it performs safely. Rollercoaster engineering is all about keeping the adrenaline rushing while minimising risks to the riders – ensuring that the excitement remains good fun.

It has been said throughout this book that engineering is all about using specialised, principled means to develop systems that meet human needs and desires. In the area of entertainment it is rather different. Engineering brings us the things that we didn't know that we wanted or needed, things that we probably would not have imagined, but now cannot live without. Engineering has thus brought an extra dimension to our lives, but some might argue that it has also brought downsides, such as addictions to technology – from spending too much time on the Internet to insatiable needs for the latest expensive gadgets. Even in the area of entertainment, engineering's affect on society is double-edged.

How society shapes engineering

> Every engineering effort is shaped by, and in turn shapes, the culture, politics, and times in which it is embedded.[3]

The discussion above might lead one to think that the products of engineering are so central to our lives that technology is one of the most fundamental forces in shaping society and its development. However, this does not mean that we are all passive free-riders in the march of technology. This view, known as 'technological determinism' is somewhat simplistic. The relationship between people and technology is far more complex than that.

Engineering and technology are as much shaped by society as society shapes them. There are always choices over avenues of research to pursue and which technologies will be developed for the market. There are also influences that are not a matter of conscious decision which will partly determine the technologies that flourish. These non-technological influences play a significant role in determining where engineers focus their efforts and which technologies become central to our lives. Once technologies are developed, the exploitation of the possibilities that they offer will be highly dependent on factors such as majority preference, accessibility of technologies in terms of cost and distribution, and even government control over elements such as broadcast technologies. For example, although it was earlier demonstrated that it is now technologically possible for those working in the knowledge economy to work remotely, there is no huge trend in this direction. Society's preference continues to be for working in industrial centres and alongside one's colleagues and thus these technological possibilities might remain dormant.

Sometimes there is a straight choice between competing technology platforms and the 'best' in engineering terms does not always win. There have been a number of battles between recording formats over recent decades. When video cassette recorders were developed there was a choice between Betamax and VHS. The rivalry between video formats could only be settled by mass agreement and Betamax eventually lost out to VHS because it was more expensive. A similar stand-off occurred when HD DVD and Blu-Ray fought for supremacy as high-definition DVD formats. Blu-Ray won out over HD DVD because of a combination of market coverage (partly by virtue of Blu-Ray players being built into Sony Playstations) and the decisions of film distribution companies to back only one of the formats. These examples show that consumer preference, often based on cost and convenience, is important in the uptake of a technology; they also show that there is a symbiotic relationship between the popularity of the technology platform and the availability of the media that it supports. In both of these cases, the popularity of the different types of video or DVD players depended significantly on the availability of films in the appropriate format. But of course the decision of film distributors to use a given format or of stores to stock it is highly dependent on the popularity of the different players. Initially the two options can co-exist, but as one technology platform proves to be more popular, or one media more widely available, the balance will tip in that direction. Thus factors not necessarily related to the quality of the technology will determine which will win out.

Many forms of engineering technology are dependent for their success on being taken up by a critical mass of users. Communications systems are obvious examples – telephones, faxes and email could only become established when enough people had adopted them to make them useful. Who wants a telephone if they don't know anyone else that they can call? And often technologies depend on infrastructure for their success, infrastructure that will only be developed if enough users show interest. For example, cars running on hydrogen fuel cells offer a low-emission and potentially (depending on how the hydrogen is produced) low carbon form of transport. But the popularity of the car is dependent on there being enough hydrogen fuel stations where the drivers can refuel. But filling stations will only offer this service if there are enough drivers to use it. Thus we have a closed circle which can only be broken by some, most likely, non-engineering consideration – a really good reason for cars with hydrogen fuel-cells to be attractive to the buyer thus creating an obvious demand for fuelling stations; or some significant commitment by a public body or large corporation to establish the infrastructure to make buying the car worthwhile.

As well as showing cases where explicit choices have been made between technologies, these examples highlight the place of market forces in determining which technologies become dominant. Engineering is about solving problems and fulfilling needs and, in the case of engineering for fun, it is about creating and meeting desires. But this suggests that there is no such thing as inherently important or significant forms of technology. Whether the work of engineers results in something worthwhile that will become part of our everyday lives, or something that will be judged a mere technical curiosity that we can live without, depends on there being a consumer for that technology. Whether that consumer is the wider public, specialist customers like the health service or Formula1 racing teams, without a market pull for the products of engineering they will die away. Engineering is about meeting human needs, and unless the need exists for a technology, or can be created, it has no real future.

Political decisions can be essential to an area of engineering research or development progressing. Take the development of nuclear power, or the building of a new nuclear power station. This is not something that an engineering research team or company can take a unilateral decision on. It is a decision that, for several reasons, has to be made by Government.

Firstly, some areas of engineering research or operation like nuclear power introduce risks to the public. As discussed in the previous chapter, although it is largely a matter of engineering concern how those risks are managed, what level of risk should be tolerated and whether perceived benefits outweigh the risk is a matter that extends beyond the technical concerns of the engineer and brings in the concerns of the public and their elected representatives. In the case of nuclear power there are also the considerations of where exactly to site a nuclear power station, or where to dispose of nuclear waste. The public interest, locally and nationally, has to be taken into consideration when making these decisions.

In addition, many engineering projects have an effect on a whole country and its infrastructure, and some centralised decisions have to be made about how they are carried out. The development of a new generation of nuclear power stations is a matter for government leadership, as it is an issue that determines the security of the energy supply for a country. Whether or when to allow the development of a new generation of stations depends on many factors, such as the capacity of the rest of the generation network, whether more coal-powered stations will be built, or whether renewable energy resources will be used to a greater extent. When engineering decisions affect a whole country, political as well as technical considerations are of paramount importance. In fact, many decisions about energy supply are dependent on international political agreements about carbon emissions and therefore the decisions about them are tied up with world politics.

Government influences technology through legislation, policies on government procurement of technologies and regulation. There are some areas of technology where it is perceived that public agreement is needed for research to continue, and this agreement comes in the form of legislation that permits the research, or rules it out. Medical research in particular, such as stem cell research or therapeutic cloning, is subject to legislation, with varying attitudes in different countries leading to different legislative decisions. In these areas, risks are weighed against benefits, and scientific evidence is assessed in the light of public feeling. Engineering tends to give rise to less controversy, but because some areas of engineering industry have an impact on the wider public close regulation is needed. Agencies such as the Environmental Protection Agency in the US and the Health and Safety Executive in the UK have particular influence and exert control over the way that engineering is practised. An area such as chemical engineering will be particularly tightly controlled by regulatory bodies, which control the use and disposal of hazardous substances. The engineering professions themselves also play a role in regulation, with institutions like the IEEE (Institution of Electrical and Electronics Engineers) in the US setting standards for new technologies to ensure reliability and interoperability.

There are some areas of engineering research and practice, however, where controversy threatens to arise, and the best way to deal with it is through dialogue between politicians, the public and the engineering community. The development of nanotechnologies, described in chapter 2, threatened to provoke a similar public outcry as was witnessed in many countries in reaction to the development of genetically modified crops. The possibility of nanomaterials having long-term polluting effects and health risks suggested that this was an area of potential public concern. In this and similar areas of emerging technology, discussion between politicians, engineers and the public is essential. Synthetic biology is just such an area. Synthetic biology is, as the name suggests, all about making biological parts by artificial means. This will involve using engineering methods to build biological material from its fundamental blocks, with the potential to engineer biological parts that are better than those supplied by nature. This area is liable to create significant public concern, since engineers could potentially be synthesising and manipulating living organisms. Decisions about the development and use of such engineering applications cannot be based only on technical considerations, wider issues of public acceptability are critical, and for all facets of the technology to be considered public debate and discussion is necessary.

Engineering: a matter of life and death?

Engineering has a profound effect on society, just as society has an influence on engineering. Does the relationship between engineering and the people whose lives it affects give rise to moral or ethical issues for engineers to deal with? Ethical issues have a high profile in medicine, but that is perhaps because medics enter into many situations where human lives are in the balance and they need to make judgements about the right course of action. Do engineers have a similar responsibility?

Engineering may not seem to give rise to the same sorts of life or death issues that spark ethical debate in medicine. However, engineering is often a life or death issue, and some of engineering's greatest achievements have cost a great deal in terms of human life. The Brooklyn Bridge caused the death or grave illness of many of those who worked on it as a result of 'the bends', or decompression sickness. The towers of the bridge were constructed by plunging caissons to the river bed, hollow boxes which were pumped free from water, so that workers could dig down to build foundations in the river bed within the confines of the caisson. Workers entered and exited the caissons through a shaft up to the water's surface, but the depth of the river bed meant that, by travelling to the surface too quickly, they suffered the effect of a rapid change in pressure, experiencing terrible body pains and extreme vomiting – a condition known as 'the bends'. As the causes of the disease had not been identified at the time, workers continued to work in the caissons, and continued to leave them swiftly after their shift ended, and continued to become ill. Even the Chief Engineer on the project, Washington Roebling, the son of John Roebling who had conceived the plan for the bridge, was struck down by the illness which incapacitated him for the rest of his life.

Other famed engineers suffered for their work. Isambard Kingdom Brunel came close to death when working on his father's great project, the Thames Tunnel. The first tunnel built for foot passengers under a navigable river, the tunnel was built by workers excavating the ground under the river. Water frequently burst into the part-built tunnel, and on one such occasion Brunel, who was working in the tunnel, came close to drowning. But he was not the only victim of the projects that he was involved with. One tunnel in his Great Western Railway (the Box Tunnel) claimed over 100 lives in its construction. There are particularly ghoulish tales surrounding the construction of his gigantic ship, the Great Eastern, with stories of a worker and his young helper being trapped between the two skins of the hull that they were riveting together. Brunel himself might be considered a martyr to his engineering work, as he died prematurely at 53 of a stroke after suffering from exhaustion from prolonged hard work.

It is, however, in the main the low-paid semi-skilled workers who die in the construction of these grand engineering projects. Should we judge the engineers who lead those projects? Or is it right to accept that progress creates some casualties? Certainly, the levels of mortality seen in the creation of Brunel's Box Tunnel would not be tolerated by today's standards.

Responsibility for the health and safety of those people that work on an engineering project lies with the engineers who are guiding the project. The Akashi Bridge in Japan, at the time of writing the longest span suspension bridge in the world, was built with no fatalities. This is as much of an engineering development over the Brooklyn Bridge as its increased span. It is also a reflection of a change in attitude across many parts of the world, where deaths in industrial accidents are no longer tolerated and significant effort is invested in preventing them.

'I only designed it!': The ethical responsibility of engineers

The engineer might be considered culpable for accident and injury caused in the construction of an engineered system, but what about its use? Is the engineer responsible for the potential uses of the technologies developed? And do engineers have responsibility for things other than human lives – does the engineer have responsibility to consider the impact of his or her work on nature and the environment? Many ethical issues arise within engineering practice. Here we will look at two areas, out of many, where dilemmas arise for the engineer.

An area of engineering practice which causes the most obvious controversy is that of defence. A great many engineers work with or within the military, or on projects that are funded by the military. The engineer might not be active on the battlefield, and may not use the weapons that they develop, but do the kinds of moral questions that apply to the justness of war apply also to engineering for defence? This thorny question will not be resolved here, but considerations relevant to the issue of defence engineering will be explored.

First, it is difficult to extricate engineering done in the interests of defence from purely civilian engineering. For example, many of the breakthroughs that support communications technologies are a result of engineering work done in the military sphere. So, condemning military engineering is not simply condemning research and development surrounding nuclear missiles, but a far wider range of research that cannot be cleanly segregated from the technologies central to our everyday lives. Of course, the communications technologies we use may have existed without the original, military-supported research, but the technology undoubtedly progressed at a swift rate because of military interest.

Many engineers working in the military sphere may feel that what they are doing helps to promote peace or at least reduce conflict. Well-armed forces might act as a deterrent to invasion and therefore prevent war. It is also the case that engineers working in the military can help design weapons so that they do what they are supposed to do – for example, crippling a country or community through damaging its infrastructure rather than taking lives, especially those of innocent civilians. Engineers can help to avoid the development of weapons like cluster bombs, which are designed to leave a wake of small bombs that will litter an area like mines – potentially causing accident or death to innocent people using that land at some future point. If engineers can design systems to work efficiently and for the exact purpose intended, surely they can help to build better weapons which significantly reduce 'collateral damage'?

However, damaging the infrastructure of a country also causes huge suffering to innocents, if electricity and clean water supply are affected. Death is not the only punishment. As we will see shortly, engineers have an agreed ethical responsibility to hold paramount the wellbeing of others, and killing through warfare, or lowering quality of life by damaging infrastructure, may seem to run strictly counter to this principle. What should an engineer do if faced with the possibility that the work he or she is engaged in will end with the loss of innocent life? The engineer might argue that if they do their job as well as possible, war can be as short and swift as possible with fewer fatalities. The engineer might focus on the civilian applications of their work. The engineer might seek to avoid working on such projects, or even to devote their professional lives to projects that help those affected by war – by rebuilding damaged infrastructure in war torn regions.

What is interesting about engineering is that these issues are very often left to the individual. The engineering profession as a whole rarely takes a stand on the justification of going to war or building weaponry to be used in war. This is somewhat unlike medicine, where the medical community often has a say and a position on the big ethical questions that affect their profession – such as the admissibility of abortion or euthanasia. Should engineering take a stand on life or death issues, or is it a matter purely of individual conscience?

Engineering practice is not afraid of tension. Just as engineers uphold the exhortation to make paramount the welfare of others yet work on weaponry, the importance of engineers' responsibility to the environment is widely accepted, yet many engineers spend their working lives developing technologies that will cause pollution that we know to be so long-lasting in its effects that it could threaten the future of a habitable earth.

The evidence is overwhelming that increases in carbon dioxide in the atmosphere, as a result of human activity, will cause significant rises in temperature. This will, in turn, cause sea levels to rise and cause changes in weather patterns which will make currently inhabited areas of the planet uninhabitable. We know that these rises in carbon dioxide are due to the burning of fossil fuels in the generation of electricity and in powering motor vehicles and aircraft. But engineers continue to design and produce such vehicles and to construct and maintain coal and gas fired power stations. Is it in any way a moral failing of engineers that they continue to develop these polluting technologies?

A few considerations are relevant to establishing the extent of engineers' culpability for designing and producing such technologies. The simple response that the engineer might make is that they only design aircraft or cars because there is a market demand for them. If anyone is at fault, it is the customers who frequently fly or the airline companies that encourage this behaviour in order to make profits. Therefore the engineer is simply meeting the desires of others. They cannot be held responsible for the desires that they have.

Of course, this is far too simplistic a position. For one thing, as the discussions at the beginning of this chapter show, the relationship between society and engineering is complex. Society develops a desire for what engineers have created just as much as engineers create what society wants or allows. So, by making powerful fuel-guzzling cars, engineers are not simply meeting a pre-existing demand, but they are stoking that demand. However, the engineer could argue that it is more than a desire on society's part to own and use cars; it has become a necessity in life. The way we live and work and the design of many of our towns and cities makes owning a car essential. Therefore the mechanical engineers who design and develop cars are meeting what has become a significant need. Engineers might argue that it is the *use* of the car that is the problem, and if too many people drive their cars half a mile to work instead of reserving them for longer journeys, that is not the fault of the engineer but of the car owner. But this is rather like the engineer arguing that they only designed weapons to be used as a deterrent, and they cannot be considered in anyway responsible if they are actually used in warfare.

To be fair, a great deal of effort in the design and development of cars, aircraft and other forms of transport is focused on making them energy efficient, or on developing vehicles that run on environmentally sustainable and less polluting fuels, such as biofuels and hydrogen fuel cells. But what about those engineers who choose to devote their career to designing fuel hungry racing cars which do not serve any of the needs that the family car does?

These questions are actually quite general questions about the extent of an engineer's responsibility for the downstream impacts of the technologies they produce, and the engineer's duty to avoid any work which is likely to cause some harm. But ethical issues arise surrounding the direct effects of engineering on the environment. As described in chapter 2, civil engineering in particular has a significant influence on the physical environment, and many civil engineering projects involve making detailed assessments of the impact of work done on the immediate environment and any species that inhabit it. How do engineers weigh the benefits of their work against any harm that might be caused to the environment? Often it comes down to a matter of weighing the benefits of what is being constructed (be that a housing complex or a reservoir) against the negative impacts on the environment (for example, loss of habitat for a particular species). But weighing costs and benefits is not easy. How does one compare the value of poor families having a home in a pleasant setting against the value of having a mating ground for a rare species? These are quite different matters that are difficult to measure on a common scale. And should we judge the natural landscape to have some *intrinsic* value, so that we might think it is inherently preferable to leave the landscape alone rather than adapting it for human good?[4]

One civil engineering project that has raised significant concerns over its impact both on the local environment and the people in it is the Three Gorges Dam which will span the Yangtzee river. The construction of the dam will allow the creation of a hydroelectric power station which will generate electricity from the power of the flowing river. This will be energy produced without releasing further carbon dioxide into the atmosphere once the construction of the dam is over, and in the long term it may well be beneficial for the global environment. But its construction means flooding 632 square kilometres of land, land that is home to many people, other animal species, and sites of cultural and archaeological interest. This controversial project brings into relief issues that arise at a smaller scale elsewhere in engineering. How do we balance the interests of a local community against wider society? And how do we balance the interests of wider society against those of the natural environment? Often engineering will create dilemmas concerning different environmental values – for example, one of the considerations that often arises in plans for wind farms in the UK is the claim that birds can be killed by flying into the blades. If birds really were threatened by the construction of wind turbines, how would we weigh the interests of indigenous species against the value of making some small contribution towards addressing the global problem of climate change?

The area of environmental ethics is well developed, and I have not really touched on the many theories and positions that are discussed in this area. But this section raises some of the questions regarding duty to the environment that an engineer needs to explore when considering the impacts of the projects that they work on and the careers that they choose.

Living by the code: ethical principles for engineers

The considerations above demonstrate the kinds of ethical dilemmas that face an engineer in deciding what to specialise in and how to practice. But all areas of engineering research and practise will involve some sort of ethical obligation, or will give rise to moral dilemmas. The fact that engineers, like medical professionals, deal with life or death situations mean that engineers will potentially be making decisions in which the health and safety of the general public, or the interests of the client are to be balanced against pressing deadlines, loyalty to one's employer, or rising costs. These practical decisions become ethical dilemmas when there is equal pressure pulling in opposite directions, and choosing either option might result in some degree of ethical culpability. What are engineers to do in these situations? Must they rely on their own personal standards?

Chapter one focused on the emergence of the professional society as a key phase in the development of engineering. The fact that engineering is considered a profession is of great significance when considering the extent of the ethical responsibility of individual engineers. Professions are characterised by the duty to be expert in an area of work, be that medicine, law or accounting, but also by the duty to uphold professional ethical principles. Professions are united not only by knowledge, but by an obligation to uphold the reputation of the profession in the public eye. And the reputation of a profession is based on open, honest, fair and ethical dealings with members of other professions and wider society.

Many professions have ethical codes that set out the rules that their members must abide by in their professional dealings. Most famous is the Hippocratic Oath, which unites doctors of today with the physicians of the distant past, in their duty to care for the ill and to avoid at all times causing people harm. This oath might be developed or augmented by the professional medical institutions of different countries in order to make it better fit the modern and local cultural or political context, and doctors' conduct is governed by the rules of the professional bodies to which they are registered, but the core duties of the physician have not altered radically.

Can the duties of the engineer be set out in a single oath or code that could be taken by all engineers? According to the author Michael Davis (in *Thinking Like an Engineer*), the first code of ethics for engineers was adopted in 1912 by the

American Institute of Electrical Engineers, with other US engineering institutions following suit in the ensuing 10 years. The major engineering institutions across the world very often promote ethical codes of conduct to their members, and many institutions may discipline or dismiss members who contravene the code of conduct.

It is worthwhile to compare two codes of conduct from different institutions in order to see the differences and similarities in the ethical principles set for their members. The current code of conduct[5] of the UK Institution of Civil engineers (ICE) is based on the following six rules:

1. All members shall discharge their professional duties with integrity.
2. All members shall only undertake work that they are competent to do.
3. All members shall have full regard for the public interest, particularly in relation to matters of health and safety, and in relation to the well-being of future generations.
4. All members shall show due regard for the environment and for the sustainable management of natural resources.
5. All members shall develop their professional knowledge, skills and competence on a continuing basis and shall give all reasonable assistance to further the education, training and continuing professional development of others.
6. All members shall:

    a. notify the Institution if convicted of a criminal offence;
    b. notify the Institution upon becoming bankrupt or disqualified as a Company Director
    c. notify the Institution of any significant breach of the Rules of Professional Conduct by another member.

These are not all purely ethical matters – for example, becoming bankrupt might not be considered by all to be a moral or ethical failing, but the majority of the rules pertain to matters of common sense morality. The US Institution of Electrical and Electronics Engineers (IEEE) has the following ten-point code of ethics:

> We, the members of the IEEE, in recognition of the importance of our technologies in affecting the quality of life throughout the world, and in accepting a personal obligation to our profession, its members and the communities we serve, do hereby commit ourselves to the highest ethical and professional conduct and agree:
>
> 1. to accept responsibility in making decisions consistent with the safety, health and welfare of the public, and to disclose promptly factors that might endanger the public or the environment;
> 2. to avoid real or perceived conflicts of interest whenever possible, and to disclose them to affected parties when they do exist;
> 3. to be honest and realistic in stating claims or estimates based on available data;
> 4. to reject bribery in all its forms;
> 5. to improve the understanding of technology, its appropriate application, and potential consequences;
> 6. to maintain and improve our technical competence and to undertake technological tasks for others only if qualified by training or experience, or after full disclosure of pertinent limitations;
> 7. to seek, accept, and offer honest criticism of technical work, to acknowledge and correct errors, and to credit properly the contributions of others;
> 8. to treat fairly all persons regardless of such factors as race, religion, gender, disability, age, or national origin;
> 9. to avoid injuring others, their property, reputation, or employment by false or malicious action;
> 10. to assist colleagues and co-workers in their professional development and to support them in following this code of ethics.

The codes are quite different in tone and level of detail, but both present a set of standards of behaviour that most would agree is a matter of common sense morality. Both codes have elements that pertain specifically to professional ethics, that is, they both refer to the engineer's duty to keep their professional knowledge up to date and to work only within that knowledge. It would be damaging for the engineering profession and would have bad consequences for the public if engineers undertook work for which they were not qualified, thus potentially creating risks for the users of the systems that they design or construct. Both put great emphasis on the importance of honesty and integrity in professional dealings – including ruling out any form of bribery (which is dealt with in the expansion of the ICE code). But what is interesting about both codes is how they differ and what they do not say.

Effort has, at times, been expended to create one code of ethics for all engineers. The Royal Academy of Engineering in the UK has developed a statement of ethical principles that applies to all engineers and it has gained significant acceptance within the community. However, it is usually used alongside the codes of conducts of the separate institutions. It is difficult to get one definitive code for all engineers, because the devil of an ethical code is in its details. Engineering codes will state that the engineer must prioritise the health and safety of the general public but exact wordings may differ and have different connotations with regard to whether the engineer must do all that is possible to make the public safe, or should do all that is reasonably practical to ensure safety. Ethical codes will bind engineers to protecting the environment – but as seen earlier, there are areas of dispute in engineering which may or may not be seen as unethical, depending on how the responsibility to the environment is understood. And codes may well be open to interpretation – how is the 'due regard' to the environment in the ICE code to be understood? It implies that it will be given consideration alongside other issues such as human wellbeing, but it does not set out what the balance should be.

These are not criticisms of the two codes, but they are features that are almost inevitable in a code of ethics. The codification of professional morality is fraught with controversy, and codes should leave room for rational reinterpretation in new situations.

Moreover, there are significant challenges in creating one code for all engineers. Engineering is now a global profession. The largest engineering firms will work on projects in all parts of the world, and will provide products and services to companies and governments internationally. But different countries are home to different cultures, and culture and morality are closely intertwined. While there are many forms of behaviour on which there is shared opinion, there are areas of ethical dispute about which different cultures may have differing views. Engineers practising in a country different from that in which they usually work and originally trained might find that their standards of acceptable behaviour are challenged, or that practises that they consider normal or acceptable might be seen as inappropriate or blameworthy. Some countries have a different attitude to the offering of payments in return for winning contracts. What should an engineer do if making such payments would count as bribery in their home country? The answer in this case is surely clear: since the practise of paying bribes can be of potential harm to the wider public (because the company who pays the highest price gets the work, not the one most qualified), the engineer has a duty to try and stamp out such practises in other countries. But other cultural differences might not be so clear-cut. Engineers in some cultures might value duty to a company more highly than duty to their profession, and might not be prepared to take a stand on practises within their company that go against professional codes. This stance might be based on a particular moral value – loyalty – and some may argue that loyalty is as important a value as the professional principles at stake.

The point is that ethical issues can never be reduced to one set of rules that clearly cover all situations. Ethical principles which guide all engineers can be identified, and they are represented in most codes of conduct. They are based on honesty, concern for the public and the environment, duty to carry out technical duties to the highest standards and so on. But these principles cannot tell the engineer what to do in all circumstances. Ethical dilemmas are hard; that is why philosophers have grappled with them for centuries. But they are an unavoidable part of the engineering profession. Engineers have to have significant technical knowledge and skill, but they must also be aware of the wider implications of their work in society. Engineers must be able to make judgements about the right course of behaviour in difficult circumstances and should be sensitive to the ethical aspects of situations they face. Therefore, the introduction of ethics teaching in engineering degrees can be of great help to engineers in dealing with these dilemmas. Engineers have a central role in society, and this brings with it a duty, and usually a desire, to do what is right. Engineers need the skills to guide them in making right decisions.

5

I tinker, therefore I am: engineering and knowledge

Because engineers work on complex problems in a complex world, they need to make use of knowledge about nature, about society and its conventions and even the way individuals' minds work. This chapter will outline the non-engineering knowledge that all engineers need to know.

But the relationship is not all one-way, with engineers only using knowledge that other disciplines create. Engineering makes a significant contribution to scientific knowledge, and to our knowledge about and understanding of the natural world. The testing of scientific theory has always involved the use of instruments in increasingly elaborate experiments, and as science has developed so has the need for complex engineering to design and construct these instruments. Engineering is even helping to develop means to better know our own minds and behaviour.

The chapter will conclude with some reflections on the nature of engineering knowledge, and what is unique about it. Although engineering makes significant use of science and mathematics and shares a great deal with them, there is something more to engineering and to what engineers know.

Beyond technology: the non-technical knowledge essential to engineering

An engineer needs to make use of a wide range of kinds of knowledge and understanding. Earlier chapters have discussed the place of science and mathematics in engineering and it should be obvious from the discussions of engineering so far that engineers cannot have any success without an understanding of relevant scientific theory and mathematical methods. But engineering is not all done in the lab or at a computer and the practical nature of engineering requires engineers to work in real world situations where there is no neat separation of one discipline from another. Engineering tasks, from the development of semiconductors for use in computers to building an oil pipeline, mark the collision of a host of different disciplines to provide a complete understanding of the matter. This means that engineers need to have, if not knowledge of all the disciplines at stake, at least an appreciation of their importance and, perhaps more importantly, engineers need to know when to call on other experts and what to do with the knowledge and advice that they can provide.

This is perhaps more a matter of skill than knowledge. Engineers use other peoples' knowledge and discoveries, and the engineering skill lies in looking in the right places and acting appropriately on the information that they acquire. This involves seeking expertise from outside the world of science and technology. Engineers may need the services of lawyers to draw up contracts or agreements for complicated construction projects, or the advice of archaeologists in order to take the correct and most cautious approach when working in areas of heritage interest. They need to use this knowledge in order to be able to work as well as possible in a complex world where many factors and facets of the local environment affect what they do. However, there are some non-technical areas of knowledge that all engineers must master to some degree to be successful.

We have seen already that engineers have to work with systems, complex artifacts and processes that involve many elements, some of which are living, autonomous human beings. A train is driven by a human driver (with the aid of a control system that might be automated to varying degrees), it has human passengers and it stops in stations full of people which are managed usually by human staff.[1] The design of the train is based on a range of engineering knowledge mostly from within mechanical and electrical engineering – knowledge of how to develop a clean and efficient engine to pull a train of the appropriate length, knowledge of how to design the train to improve its aerodynamic properties, and knowledge of how to introduce systems to reduce accidents through driver error. It is this last example that makes the need for knowledge of how people behave apparent. To design the control systems of a train there should be some understanding of how a driver will operate the train, what kinds of controls are easy to use and which are less intuitive – either hard to master or tending to lead to mistakes. To a large extent, passenger vehicles like trains and trams which run on fixed routes are becoming increasingly automated, with the driver acting more as a safety device than an operator – that is, they are there to act when things go wrong, rather than to make sure that things are going right. But the design of these automated vehicles still requires an understanding of the role of the driver and what their behaviour and thought processes are likely to be. If a train cab becomes so automated that the driver becomes catatonic with boredom and unable to act in an emergency, the level of automation becomes counter-productive. This has to be considered when engineered systems are designed for human interaction.

The operator is not the only 'human factor' in an engineered system. Trains designed for passenger travel must be developed with the comfort and convenience of customers in mind (though a lot of commuters may not be easily persuaded that this is frequently achieved). One of the crucial aspects of psychology for the engineer is knowledge of how passengers will behave in an accident – designing a safe train is dependent on knowing what people tend to do when things go wrong. 'Human factors' involves the study of this kind of behaviour, and may be used by the engineer to make sure that there are minimal obstacles in reasonable escape routes, or alternatively, to design obstacles in exit routes to make dangerous or obstructive behaviour less likely. For example, in the evacuation of an aircraft, logjams are not only caused by people being unable to open the emergency doors but also by the tendency of people to trample over others to get out first. Engineers need to consider human behaviour in this kind of situation and design to control it, as far as that is possible.

Engineers need to understand people because engineered artifacts are supposedly created to meet a human need and this cannot be achieved successfully without an understanding of those needs. But it is somewhat disingenuous to suggest that engineers are actually always good at this. Most people have encountered unintuitive technologies that are not controlled in an obvious way. Technologies are often not designed with users in mind, and it is often the user that has to learn and to adapt in order to fit the technology. This can often have serious downsides. The computer mouse is often cited as such an example, as its use is uncomfortable for many and can engender bad posture and lead to repetitive strain injury. This is not surprising since the mouse was originally designed in 1963 (by Douglas Englebart) long before people began to spend all of their working day in front of a computer operating a mouse or, indeed, much of their leisure time doing the same. The key to better engineering design is better understanding of peoples' physiology and psychology – and that should include the whole range of likely users. The male engineering graduate student may interact with technologies in a different way to the retired female nurse.

The last chapter showed that there is a range of non-technological factors that determine whether a product of engineering will be successful or not. Engineers should not consider themselves passive players in this game, waiting to see if the conditions are right for their invention to flourish. Instead, they need to acquire the skills and experience to allow them to develop products that are more likely to grab a slice of the market. This involves an understanding of what is likely to appeal to customers (so an appreciation of psychology also comes in handy here) and what there is likely to be a market for at a given time. It also involves an appreciation of how to make products at competitive costs. Turning an idea into a marketable product involves finding a way of making that product so that it is affordable – the most ingenious piece of technology in the world will not gain a foothold if it is too expensive or difficult to make. This applies equally to large global markets and to goods produced for specific communities. An engineer might be designing water purification or sanitation devices intended for remote communities with little money to spend; success depends on an understanding of how the needs and resources of these communities differ from others.[2]

Engineering innovation is very often supported by business innovation. The story of how Google succeeded in turning a mathematical process into an engineered system has already been told, in part. But being able to turn a neat application of a mathematical function into a process that worked at scale was not the only thing that made Google successful. What accounts for their huge impact was working out how to make money out of it, so that Brin and Page could attract investors and create a real business with potential for growth. The way that they make money is as innovative as the way the search works. The key issue was not to compromise the search results – no one should be able to pay to get to the top. So they added sponsored links to the page that would be returned independently of the search results, yet linked to the results. This was done by getting companies to bid to be one of the sponsored links that appears when certain search words are hit. Some searches are more valuable than others – eg 'hotels' will be worth a lot, 'thermodynamics' less so. I know that 'thermodynamics' is not perceived to be worth much, because when I type it into the search bar, no sponsored links appear. This is most likely because people googling 'hotels' are likely to be looking for online booking sites, whereas people searching for 'thermodynamics' are likely to be looking simply for more information on the subject. The process of bidding for advertising space on given search results was not done by Google first, but they identified this as the way to make money out of their search process, and it delivered them a business model that allowed Google to make money while still fulfilling its technical function. This kind of business insight is critical in turning a good idea into a successful product. There are a lot of solutions to every problem that an engineer can grapple with, and business viability is a key factor in deciding which ones will succeed.

Global and local politics play a huge role in engineering, from international energy supply agreements to planning permission for a skyscraper. Again, the engineer can best control these factors by understanding them. The importance of political knowledge is nicely illustrated by looking at the construction of a particular oil pipeline and the political issues that determined how it was built.[3]

The Baku-Tbilisi-Ceyhan (BTC) pipeline carries oil from the Caspian Sea to the Mediterranean, crossing three countries, and was built by a consortium of companies headed by BP (British Petroleum). It begins at Sangachal, near Baku, the capital of Azerbaijan, travels through Georgia via Tbilisi and reaches its terminus at Ceyhan in Turkey. At the time of the development of the pipeline, a number of political issues constrained the possible route. Azerbaijan is flanked on one side by Russia. The relationship between these two countries has for many decades been hostile, dating right back to the Armenian-Azerbaijani war which broke out after the Russian Revolution. Azerbaijan's other neighbour is Iran, and relationships between Iran and the US were tense, making a route through Iran impossible given the involvement of American companies in the project. Hence the route out of Azerbaijan was through Georgia, though Georgian politicians made stipulations about exactly where the pipe could be laid.

The planning of the pipeline route was thus fraught with political obstacles. These were just as important in considering the route as geographical and economic concerns. Understanding these issues was not something that the engineers could leave to others. Many engineering projects play out these issues on the smaller scale, as they involve the interests of different stakeholders – the customer for which a system is built, the wider community, special interest groups like environmental action groups and so on. Engineers cannot and should not consider dismissing these factors as outside of the concerns of an engineer. If engineering is about improving quality of human life, the interests of humans cannot be dismissed in the engineering process.

The quest for knowledge: engineering in support of science

A central aim of science is surely the development of ever more detailed knowledge of the natural world, and engineering plays a major role in supporting the growth of scientific knowledge. There are a number of engineered technologies that have proved beneficial to many different branches of science, or which have had useful and unexpected implications for scientific discovery. The scanning electron microscope was developed for commercial use by engineers in Cambridge, and is now an essential tool for researchers working across the scientific disciplines. Leaps forward in computing speed and power have enabled increasingly complex data processing, calculations and predictions. It was the opportunities created by computing capacity that turned the sequencing of the human genome from a theoretical possibility into a reality, thus enabling further research into genomics which would have been impossible without the tools of engineering.

Very often, engineers and scientists work together as part of a team focused on a particular research project. CERN (European Organisation for Nuclear Research) in Geneva is a hot-bed of collaboration between experimental physicists and engineers. CERN is home to an array of experimental teams working on different projects, but one of its most high profile concerns is the Large Hadron Collider (LHC), which was first switched into action in September 2008. The collider was designed to recreate conditions a billionth of a second after the big bang, in order to allow detailed and extended study of them. One of the main aims is to solve the mystery of how particles have mass, and to identify the constituents of the hitherto mysterious matter that forms the vast majority of the universe.

The engineering challenges created by the LHC are in many ways classic engineering puzzles – but they arise in a very unusual context and with a very unusual purpose. The collider comprises a 27 km circular tunnel in which particles are accelerated before passing through four separate detectors. One of the most obvious engineering challenges posed by its construction was due to the fact that this highly sensitive equipment had to be housed underground, in a well-protected environment. The particles in the collider are accelerated by the influence of superconducting magnets, which only function at near absolute zero. Creating this protective environment where the critical temperatures can be maintained was essential for the reliability of the equipment, and ensuring this reliability was the central challenge for the engineers involved. The experiments carried out in the LHC cannot be stopped and started at whim, with small glitches tweaked before switching the system back into action. Because it contains parts that cannot be accessed without shutting down the system for months, the system has to be engineered so that it will function reliably once set into action. Just as satellites are set out on their mission with the need for a guarantee that they will continue to function throughout their lonely missions, the collider, invested in by many countries and supporting the work of thousands of scientists, had to be built to work reliably.

Shortly after it was first activated, the collider developed faults that caused it to halt its work for months. The faults might be frustrating (and no doubt embarrassing) for the engineers and scientists involved, but it was a foreseeable outcome for such a complex and sensitive system. The experience of the LHC engineers actually demonstrates the criticality of engineers' role in the project, and shows that large experiments are as tough a test of the work of engineers as they are of the work of theoreticians. Big science opens up territory that is uncharted for both scientists and engineers.

The LHC is not the world's only particle accelerator, however. In fact, there are many particle accelerators used, not as colliders, but for the creation of intense beams of light that have applications in a wide variety of experiments. Light beams created by synchrotrons (particle accelerators) can be used to reveal the structure of matter by shining those beams through samples of material. By offering a way to observe the structure of various materials, they have applications in a vast range of scientific fields including materials science, biology and even

Figure 4 Diamond Light Source, the UK's national synchrotron facility, under construction in April 2004, on the Harwell Science and Innovation Campus, south Oxfordshire, UK. *Courtesy of Diamond Light Source Ltd*

medicine. Although their functioning is based on a theoretical understanding of the behaviour and properties of fundamental particles, the construction and operation of these facilities is the task of engineers. Experiments are devised by scientists to answer the questions thrown up by their theories, but the process of answering the questions raised by modern science is impossible without the feats of engineering.

The role of engineering in supporting theoretical science has a long history. In the thirteenth century the development of astronomy created the need for ever more precise measuring devices for plotting the movement of celestial bodies. Precision engineering is an area of mechanical engineering focused on the production of machine parts and other artifacts to high levels of exactitude and uniformity, and it has roots in the art of making highly accurate measuring instruments. Precision engineering is an area where engineering progress is easy to observe by way of the mind-blowing advances in accuracy. A major leap forward in this area was the creation by John Wilkinson in 1774 of a boring machine capable of making a hole with a diameter of 1270 mm with an error of 1mm (which was essential for the development of an efficient steam engine). Now it is possible to machine with levels of accuracy defined at the nanoscale. That is to say, it is possible to work with levels of accuracy of 100 nanometres or less, where a nanometre is one billionth of a metre. This advance in precision engineering has allowed it to support continuing advances in sciences such as cosmology.

Gravity probe B, which was launched in April 2004 was devised to measure the curvature of spacetime, in order to test the predictions made by Einstein's General Theory of Relativity. To do this, the probe needed highly accurate gyroscopes which would detect the change in orientation of the probe, revealing the degree of curvature of the region of space-time in which it was travelling. The kinds of changes in direction that are predicted are extremely small, and the gyroscopes had to be highly sensitive. Central to acheiving this sensitivity were the extremely smooth spherical rotors that form part of the gyroscope. The further they diverge from smoothness, the less accurate the results of the gravity probe. Made from quartz, the spheres are polished to create homogenous 1.5 inch spheres, which deviate from perfect smoothness by a few layers of atoms.

This level of accuracy is reflective of the standards in modern precision engineering, where highly precise and sensitive components are made for high-tech instruments from cameras to computers. Through the development of engineering skills that have a direct link to the work of craftsmen-instrument makers, cosmology and astronomy have been able to develop into sciences that can really reveal the fundamental properties of our universe. These examples, from the civil engineering involved in excavating land for a particle accelerator, to the development of high-speed computing, show that engineering is not just a user of science but is in fact a facilitator of science. A great deal of experimental work done today is arguably at least as dependent on engineering as engineering is on the work of theoretical and experimental science.

Engineers do not only make things for science to use, they can also lend their methods to other areas of science. Systems biology is an area of research that offers the possibility of a revolutionary approach to healthcare. Much of medical science is based on understanding processes in individual areas of the body, diseases affecting a single organ or type of cell. Our understanding of the human body has been influenced hugely over the last century by the rise of molecular biology – understanding life through the properties of its fundamental building blocks. But humans are complex systems, and there is a clear need for understanding how the body functions as a whole, how different organs interact, and how processes at the molecular level influence those at the cellular level or level of the organ, and vice versa.

Engineers have developed an increasingly sophisticated understanding of complex systems, and the techniques developed for describing and modelling such systems can potentially play a huge role in biology and medicine. The mathematical techniques exploited by engineers will in particular have a significant role to play in mapping the complex interactions between processes going on at all levels of organisation of the human body. Since the sequencing of the human genome we have a Pandora's box of information that we can use to better understand the biological characteristics of individuals, and this can be used to understand how treatments can be designed to better match the individual and also to understand the propensities of individuals to particular illnesses, offering the hope for proactive prevention of disease. With the continual growth of processing speed and capacity for computer memory, we have the hope of being able to use the massive potential banks of information about each of our bodies, to better understand how they work and how they might fail.

Sometimes engineers have to become like scientists and develop their own scientific theories. Engineering problems may present themselves which require direct investigation without the aid of pre-existing science. Claude Shannon, an electrical engineer responsible for significant innovations in communications technologies, is credited with the invention of the applied science of information theory. The application of information theory is in finding a method of transmitting information over a channel where there is interference. His problem was finding a means for transmitting information so that it can be decoded at its destination even when some of it is lost in transit. He dealt with this problem by investigating the nature of communication and gained inspiration from studying the way that humans communicate using ordinary language. He noted that we can usually understand the gist of what is being said even when we do not catch every word of a sentence, and he used this insight to develop ways of compressing data for transmission. Shannon's work helped to establish the discipline of information theory, solve an engineering problem, and highlight interesting aspects of the way humans communicate. The area of biomimetics, where engineers copy mechanisms found in nature in the development of technologies, is another example of engineers being like scientists and engaging in direct investigation of the natural world. By this means engineers can shed light on phenomena of interest to science while solving their own technical problems.

Philosophical engineering: knowing our minds

Given that engineering is focused on the development of products and systems with practical outcomes, it might seem unlikely that engineers can collaborate fruitfully with philosophers. But in fact the products of engineering, and even the way that engineers work, can offer enlightenment on some of the enduring questions of philosophy. Engineering method and specific engineered tools can throw light on some philosophical and sociological questions about how our minds work.

For example, philosophers worry about how we can know that there is a real world out there and what it is like. They wonder how we can be sure that we are not just dreaming the world around us, or that it is not just some elaborate virtual reality created by a manipulative experimenter. But when philosophers ask this question they often think of knowledge in terms of the things we know from perceptual experience, from observing the world around us. Their concern is how we can know that our experiences are real and that the factual knowledge derived from those experiences is true. When philosophers raise such questions, they are assuming there is some kind of gap or bridge to be crossed between the world we observe (appearance) and the way the world is (reality). But this gap is not obvious when it comes to practical knowledge, especially the highly developed skills of engineers.

For example, philosophers worry about how we can know that other people have minds, and therefore inner experiences like the 'private' inner experiences we have. How do we know they are not mere robots, programmed to behave like us but without a conscious inner life? This doubt arises because there is seen to be a gap to be filled between physical actions and the inner thoughts that somehow mysteriously cause outward behaviour. Neuroscience has done a lot to close that gap, by describing in close detail the processes in the brain that bring about physical behaviour. But biomedical engineering, the discipline that has created devices like the i-Limb hand described in chapter 2, has closed the gap even further. The creation of artificial limbs that respond to an individual's intentions gives a different perspective into the relation between mind and body. The design and fitting of such limbs involves intervening in the processes which connect mental events like thoughts and intentions with physical behaviour. If we can build physical devices that are made to function by a patient's thoughts, have we not succeeded in controlling the relationship between the mental and the physical? Being able to work in this way surely narrows the window for philosophical doubt about the intimate connection between the inner mind and outer physical behaviour.

Robotics is another area of engineering where there are explorations going on at the edge of philosophy and psychology. Much of robotics might be focused on the development of tools to perform tasks that are very repetitious (assembly line robots in factories) or demand great accuracy (robotic surgeons), but there is an area of robotics that interfaces with the study of Artificial Intelligence (AI). In this area there is great interest in replicating the behaviour, decision making processes and even demeanour of humans. This involves understanding the nature of human intelligence and even the nature of consciousness, and collaborations between philosophers and engineers have proved stimulating in this area. But it may not be just a matter of engineers listening to philosophers' views. As engineers get better at developing robotic systems that display activity approximating what we might consider to be intelligent behaviour, or behaviour indicative of consciousness, they might have lessons for the philosopher. A great deal can be known about a system through how it is put together, and if engineers can succeed in constructing a system that passes as intelligent or conscious, they will possess secrets about the nature of these mysterious attributes that no amount of pure philosophical theorising can achieve.[4]

The products of software engineering have also provided opportunities for discovery in the social sciences, by creating laboratories in which human behaviour can be observed, understood, and even manipulated. CAVEs (Cave Automatic Virtual Environments) create a virtual environment which can be designed and adapted to observe the details of human behaviour in different contexts. An image is projected on to the walls of a cube shaped room, which allow a subject to experience a 3-dimensional recreation of a space of interest. This makes it possible for experimenters to observe peoples' behaviour in a variety of settings. This might be as simple as seeing how people walk around and interact with products in supermarkets, which is useful in deciding where to place the most high value products on supermarket shelves; or it might involve observing people's behaviour in situations such as evacuation of spaces in the case of emergency. These studies can then be used to design spaces in order to capitalise on natural human behaviour – be that getting someone to pick up your brand of chocolate, or designing emergency exits in a building so that they are easy to use and don't get congested. This is, of course, knowledge that is just as useful to engineers as it is to psychologists. So even in the 'soft' social or psychological sciences, engineering is not just making use of knowledge, it is helping to create it.

What is unique about engineering knowledge?

Having looked at the interface of engineering with various disciplines in which specialised knowledge is developed, it is worth asking what it is that engineers know that experts in other disciplines do not. Is engineering knowledge different to the knowledge created in other disciplines? Does engineering produce knowledge at all, or does it only use knowledge in order to produce artifacts?

Engineering is clearly inherently practical rather than theoretical and its main focus is on the development of things and not facts – in the form of reliable systems, products and processes. Engineers succeed in this aim by using their ability to design, test and make what is required. The key idea here is abilities or skills; these are the essence of engineering knowledge. While there is no doubt much skill involved in being a successful experimental scientist or a mathematical physicist, knowledge is their main product and tool. But engineers, while having a great deal of mathematical and scientific knowledge, also have an ability to solve the problems they encounter. They have the ability to figure out what kind of solution is needed, how to design such a solution and how to translate that design into a physical object – whether they do that themselves or direct others to do it. We might say that what engineers have is *know-how*.

Philosophers examining the nature of knowledge contrast 'knowing how' with 'knowing that'. I know that chlorophyll makes plants green, that the atomic number for gold is 79 and that copper is a good conductor of electricity. I can check these facts with others and I can write them down and pass them on. However, I also know how to ride a bicycle, how to adjust the brakes on a bicycle and how to put a bicycle together; these are not facts that I can recall, but skills that I can exercise. I can explain to someone else how to do these things – but only to a point. I would have a hard job explaining how to ride a bicycle for example, and although I could go some way to describing how to adjust the brakes on a bicycle, a person reading my description would not pass as knowing how to adjust the brakes on a bike. They only have this knowledge if they can do it correctly, if they can exercise the knowledge. Acquisition of this kind of knowledge generally only comes about through practice. The same goes for building a bicycle – a lot of instructions can be written down but merely reading those instructions does not pass on the ability. I can read a book about building a bicycle but you are unlikely to judge me as knowing how to build a bike unless I am able to do so successfully – which I can only demonstrate by doing it.

Engineering is built on and crucially involves knowledge in the forms of such skills. Knowing how to tighten a nut just to the right point, without turning it too far, is a skill more or less ineffable but based on experience and on learning what feels right. Using a spanner to tighten a nut is not engineering, but many products of engineering depend for their success on people possessing just this implicit knowledge. A bridge design will be developed with the use of mathematical method and calculations, but the safety and functionality is also crucially dependent on the tacit knowledge of those who executed the construction of the bridge.

It is worth making an aside here about the role of technicians in engineering. A distinction is usually made between the engineer who designs and manages projects and the technicians who carry out specified practical tasks involved in manufacturing, project delivery and product maintenance. In the UK in particular, this distinction is somewhat blurred, especially due to the fact that technicians often call themselves engineers – for example, the technicians who fix broken lifts might be advertised as 'lift engineers'. There is sometimes concern among the engineering community that it is misleading and even harmful for the name 'engineer' to be used by technicians. It is felt that the average member of the public identifies engineering with the work of technicians, which involves hands-on fixing of bits of machinery (the links between the word 'engineer' and 'engine' probably do not help here).

However, the technicians that help to realise the plans of engineering designers and to maintain engineered infrastructure are essential to engineering. This has been so since the industrial revolution. In her book *Men of Iron*, Sally Dugan says of the Brunels' colleague Henry Maudslay: 'Maudslay had craft skills that could not be learnt from books, and without his technical know-how – passed on by word of mouth to his apprentices – both of the Brunels' designs would have remained nothing more than marks on paper.' This is not just a matter of Maudslay simply building what was planned on paper by the Brunels, and realising their vision by making it concrete. The kinds of designs that they were producing were novel, and building them was not simple. It demanded knowledge of how to build effective and high quality machinery and how to adapt that knowledge to the creation of new designs. This involved an appreciation of the importance of constructing machines from uniform, interchangeable parts but also the ability to build machines that are capable of manufacturing these interchangeable parts. Maudslay was expert at making machine tools, and that expertise came not from learning theories, but through apprenticeship: learning by experience.

Maudslay counts amongst the ranks of the great engineers, despite his experience being rooted in the crafts (he developed his skills by working as a lock maker). But the technicians who carry out everyday construction and maintenance tasks have played a central role in the growth of engineering. It was almost entirely people from these ranks who lost their lives to engineering projects, and although the scale of loss in projects like the Hoover Dam (which claimed 114 workers) is no longer tolerated in developed countries at least, technicians continue to contribute hard toil to deliver great engineering designs. The Bird's Nest Stadium which was the centrepiece of the Beijing Olympics in 2008 was turned from an architectural vision to an engineering reality by engineers at Arup. These engineers used the analytic tools at their disposal to create a plan for constructing the complex woven structure that forms the outside of the stadium. But it was hours of gruelling work by welders working at height and in awkward positions that helped to convert these plans into the physical stadium. The skill and know-how of engineering technicians is therefore a crucial part of engineering knowledge. It is knowledge gained only by working on engineering projects, and it is knowledge that is essential for those projects to be completed.

Many writers on engineering emphasise the importance of skills learned through experience over explicitly stated knowledge. James L. Adams, in his book *Flying Buttresses, Entropy and O Rings* writes: 'Some engineers are difficult to distinguish from mathematicians and scientists. They work with esoteric and sophisticated theory and powerful computers. Others are very close to technicians. They have a pragmatic hands-on approach and an almost mystical sense of rightness, working by feeling and instinct based on experience.' For decisions made even at the level of design and project management are often dependent on tacit, practical knowledge. It takes a great deal of engineering experience to be able to foresee and diagnose failures in a system – the uncertainty associated with potential modes of failure means that it is never possible to simply list potential failure modes and thus learn to identify them by rote learning of a set of rules. Engineers have to deal with situations that are always subtly different from those they have previously encountered – different clients, procuring projects based in different geographical situations and so on. Hence, the experience engineers acquire cannot simply be reduced to a set of rules for decision-making. Much engineering knowledge is thus related to the basic kind of know-how described above, a sometimes painfully acquired ability to sense when something is right or not.

It is know-how, then, that makes engineering knowledge unique. Engineers differ from scientists in that they not only have an understanding of mathematics and physical theory, but they also know how to apply that understanding. They can think in a rigorous and systematic way just like scientific theory-builders, but they can also use their systems-thinking to create systems that serve a particular purpose. Of course, this is not to say that all engineering knowledge consists of ineffable skills that one engineer cannot convey to another. The progress of engineering from the status of a craft is due to the development of sophisticated methods that can be codified and therefore shared and taught. However, an engineer is not made by reading about the lessons learned by others. To be an engineer is to practice engineering and to develop practical engineering knowledge.

6 Legacy and inheritance: engineering's past and future

This book will close with a look forward to the future challenges for engineering. This is not a survey of the 'cutting edge' technologies that are about to transform the future of engineering. The pace of change in engineering and technology, and the many influences that shape engineering development, mean that it would not be long before any predictions became old news or nonsense. Instead, this short chapter will look at the challenges that engineering faces, either as a result of the legacy it has left for itself, or because of the way that we have used the products of engineering.

Preceding chapters have shown how in developed countries, and increasingly in the developing world, engineering has shaped the landscape and people's ways of life. However, the resulting changes in the local and global environments have created a whole new range of problems for engineering to address. Firstly, a significant portion of the world's population is dependent on engineered infrastructure to lead their everyday lives – from the transport networks to communications systems. This means that there is a significant task for engineers to maintain that infrastructure so it does not crack under the strain of constant use. Engineers also have to be alert to the external threats to the infrastructure and strategies are needed for defending it from attack. Secondly, the work of engineers has had undeniably negative consequences. A great deal of engineering activity has resulted in pollution on both local and global scales. Increases in the levels of carbon dioxide in Earth's atmosphere as a result of human activity threaten climatic patterns and could have a huge impact on human life. And the engineering activities that have contributed to that increase in carbon dioxide – power generation, car travel, air travel, the use of ever more modern, power-hungry devices – are likely to grow as the developing world seeks to provide the comforts of modern life for its people. Although these negative impacts are not the 'fault' of engineers as such, but are the result of human behaviour across large regions of the world, it is undeniable that engineered technologies produce a good proportion of the world's greenhouse gasses.

In areas of the world where engineering has already changed human life, the challenge for engineers is to maintain the engineering infrastructure to preserve quality of life and to develop it in such a way that it does not continue to pollute the global environment. In other parts of the world, the challenge is to create a sustainable infrastructure that poses the least damage to local and global environments. This chapter will set out what these challenges mean for engineering.

The ageing infrastructure

The engineered infrastructure faces many threats that engineers must protect it against. In countries like the UK, the engineered networks that support our lives are almost always functioning at near-capacity, at just the right level to provide the services needed. An efficient infrastructure is a good thing, but it is important that there is spare capacity in order to deal with out of the ordinary events. Unfortunately, population growth and the concentration of population in city areas mean that this spare capacity is constantly eaten away.

In the US and Europe there was a huge boom in infrastructure development in the nineteenth century. Road and rail networks sprang up rapidly, and water and sewage systems were rebuilt to meet the needs of modern towns and cities. Most countries have experienced periods of rapid housing development designed to meet the needs of the population of the time. All of these aspects of the infrastructure are subject to a range of threats.

The passage of time is one of the most obvious problems. Engineered systems have a more or less fixed life expectancy – nothing can be built to last forever. But the products of engineering are so embedded in our towns and cities that maintenance and replacement is an ongoing headache. Water mains and sewers that supply water to and remove waste water from homes and businesses suffer with age and use, but the process of updating them can mean massive disruption, digging up busy roads and interrupting the functioning of the transport network. But these are essential services that cannot be risked, and so intrusive work has to be carried out to keep them functioning.

As time moves on, changes in population have to be catered for; systems built for a population of a given size often end up carrying the burden of a much larger body of users. Transport systems are extremely vulnerable to growth in the numbers of users, with more commuters squeezed onto trains, and more drivers causing gridlock on the road system. Where it is not possible to build extra capacity, these systems have to be carefully managed to meet the demands made on them. The UK road system is now watched over by thousands of cameras, linked to a nerve centre where traffic is monitored and decisions are made to divert or stop the flow of traffic when accidents occur or congestion builds up. Although development of the highway system itself is essential, building intelligent systems that make the most of existing capacity is extremely important. Potential lies in the global positioning systems (GPS) that many drivers use for navigation to help them make choices to improve their journeys and those of others. By capturing and using information about the free capacity on various roads, traffic can be managed better by the drivers themselves. Road charges can be introduced and used to discourage journeys at busy times, or to make certain routes more desirable by making them cheaper to use. As our transport networks get busier, they need to get smarter. The users too need to play their part. Drivers are relatively free to make their own choices about the times that they travel and the routes they take. Systems that relay information about the state of the roads in order to manage them better will only work if people act on that information. And that depends on a little altruism amongst road users – at least until the development of the autonomous car that makes its own decisions about routes and diversions, based on the information relayed to it.

The effects of climate change are posing a significant problem for the built infrastructure. Severe heat in the summer can create a demand for air-conditioning that electricity networks cannot keep up with and that adds to the climate problem by increasing energy use. Brick built houses constructed in the UK at the turn of the twentieth century were not built to insulate against either heat or cold, and need to be heated and cooled in extremes of temperature. This creates a vicious cycle that exacerbates the effects of climate change. Across the world climate change is affecting weather patterns and it is likely to lead to more frequent episodes of extreme weather. These events, alongside changes in population, are having a huge impact on the abilities of towns and cities to deal with waste water. The devastating effects of hurricane Katrina in 2005 show the potential for destruction when flood defences are overwhelmed. The summer of 2007 saw floods across the UK that tore through peoples' homes and caused failure of electricity and water distribution systems. This was due in part to the fact that storm drains were not able to cope with the amounts of water flowing into them, as they were not designed to deal with such hitherto rare flash-flooding events. Waste water systems need to be redeveloped to deal with weather patterns quite different to those foreseen when they were built.

The efficacy of such systems also depends on other, apparently unconnected, elements in the infrastructure. As more of the landscape is paved over or covered in concrete – be it roads, car parks, or peoples' driveways – the harder it is for rainwater to soak into the ground. Instead it runs rapidly off these hard surfaces into waste water systems that cannot cope with the surge. Innovative solutions to this problem include laying more 'green roofs' – that is, covering the roofs of large buildings with lawns of grass or sedum. Rainwater will seep through these natural roof coverings and run into gutters at a much slower rate than it does from hard roofs. They also have the advantage of providing insulation for the buildings they cover. This is a good example of the complexity of infrastructure systems – taken as a whole, changes in one area of a country's infrastructure can have effects on another, for good or bad, that are difficult to predict.

Terrorism is also a potential threat to infrastructure throughout the world. As terrorist attacks become more sophisticated, they have the capacity to damage critical infrastructure causing huge disruption. For example, even disturbing the GPS systems in a country (which might not seem that fundamental to providing our daily needs) could cause chaos, with ambulances and emergency services unable to get to where they are needed. The reliance that we have developed on the technologically-driven infrastructure makes us more vulnerable to the impacts of an attack. The engineer's task is to try to design systems to be as reliable as possible.

But reliability is hard to engineer when the transport, utilities and communications networks have to meet needs that their designers may not have foreseen. Engineers have to think far into the future, to create systems that will serve not only current needs but will be resilient to changes in demand. Such foresight is difficult to develop, but research into potential future scenarios is a project for both engineers and politicians to invest in. Without straying into science fiction, experts in public policy and public infrastructure can and must actively identify potential future problems, and plan and design for them. Engineering projects deal in timescales that cannot be dictated by short term political or business goals, and one of the great future challenges for engineering is the challenge of seeing into the future – including those dark corners where the unexpected hides.

Building engineering capacity

Engineers have, in many ways, succeeded in their aim to improve the human experience for people in the developed world. We can enjoy reliable heat, light and water supplies. We can travel quickly and easily in our own vehicles or on public transport and overseas travel is cheap and accessible to many. Developing countries, understandably and justifiably, are now striving for the same standards of living. This means that fast-growing countries like India and China are seeing rapid increases in the number of cars on the road and exponential expansion in electricity generation and the building of power stations. The relentless march of construction means that many more square miles of the planet are being covered and the process of achieving this is power hungry, resource hungry and polluting. All of this is adding to global carbon emissions and threatening any attempts by engineers to make up for the damage done in the last century and a half in the west.

Can we argue that this is wrong, that it should be called to a halt? Should developing countries be made to implement the cleaner, more efficient technologies that engineers are now seeking to design? Should the opportunity be taken to work with a new model of power generation, one that is sustainable and clean? The arguments are compelling for this position, but is it fair to hamper the progress of developing countries by demanding that they implement technologies that are expensive and that are still in their infancy? Could any country ever dictate to another that it should rely on currently unreliable and expensive renewable energy sources? These are hard questions to answer. The rapid pace of change in developing countries depends to a large extent on their ability to adopt tried and tested technologies. This allows for rapid expansion, and whatever that rapid expansion may bring, it is lifting millions of people out of poverty and improving their quality of life. But this improved quality of life will be short lived if the result is changes in climate that render large areas of the world uninhabitable. A compromise has to be reached.

Engineers worldwide can help support developing countries by investing in the project of finding clean, efficient and renewable methods of developing electricity. This is a global problem which, if addressed, can help developed and developing countries alike. A discussion of the opportunities and challenges for energy generation follows below.

A promising strategy is to find ways of devising technologies for developing countries without following the trajectory of engineering development in the west. African countries, for example, are quite different to countries in Europe or North America, in terms of size, population density, and opportunities for exploiting renewable energy. This calls for innovative responses to the engineering challenges that they face. For example, in the western world communications networks were created by laying cables and stringing telephone wires across neighbourhoods, methods which are not suitable for countries such as Africa with small, remote communities. But Africa has been able to exploit mobile technologies in a way that has allowed them to leapfrog the early stages of communication networks in the UK, using shared telephones to keep remote communities in contact and using personal cellphones to provide services such as financial transactions – from transferring money to paying for bus travel. Using the capability of available technologies a very different financial and communications infrastructure has been developed.

The opportunity to develop infrastructure in a novel and sustainable way also presents itself in countries overcoming damage caused by conflict and natural disasters. Engineers can play a vital role in a country's renewal provided real investment is made into rebuilding infrastructure. For example, in post-conflict Afghanistan, roads and bridges were rapidly rebuilt but they were not planned effectively, so that erosion of hastily constructed roads and bridge collapses caused further problems.[1] Engineers are good at learning lessons from past failures, and a global effort has to be made to ensure that infrastructure is built or rebuilt in the developing world to be innovative and sustainable.

STARTING FROM SCRATCH? ARUP AND THE ECO-CITY

Could it be possible to create cities that work in completely different ways to those in the developed world – cities which have a much smaller carbon footprint whilst proffering a high quality of life for their inhabitants? The city of Dongtan, planned for Chongming Island close to Shanghai, was an experiment to test the

STARTING FROM SCRATCH? ARUP AND THE ECO-CITY (*cont.*)

possibility of creating a city that will be completely sustainable. The Shanghai Industrial Investment Corporation (SIIC) made the decision to develop Dongtan on an island in the Yangtze, and enlisted the engineering firm Arup to create a masterplan for the city. The aim was that Dongtan would be powered by energy generated locally from a mixture of renewable sources including wind-power, solar-power, and CHP (combined heat and power plant) that runs on biomass and recycled waste. Dongtan was designed to be as energy-efficient as possible so that it could be self-sustaining in terms of its energy. Public transport would be provided by zero-emissions vehicles powered by hydrogen or batteries, but the city was designed to make walking and cycling as easy and attractive as possible.

Dongtan would provide as much food as possible for its inhabitants, and water supplies would be fed by collected rainwater and recycled water. Waste would be re-used as far as possible and recycled, or used for energy generation. The main strategy for meeting these high aims was to create a holistic plan for the city, making the most of existing technologies.

Building a city in this way is no doubt the easiest way to create a city that is sustainable and efficient, and will not suffer from the legacy of ageing infrastructure and old, energy inefficient buildings that form the body of many cities.

Unfortunately, starting from scratch is not possible for most cities in the world. In these cases engineering vision will have to focus on improving what already exists. Although the Dongtan eco-city is an admirable idea, the reality might be that sustainability can only be achieved piecemeal, as part of an ongoing programme of infrastructure modernisation and improvement.

Figure 5 Artist's impression of Dongtan eco-city, designed by Arup. *Photo © Arup*

Energy challenges

Although the evidence for climate change and its human cause has been accepted with overwhelming consensus amongst scientists, the phenomenon is nevertheless riddled with uncertainty concerning the exact impacts of rising temperatures. But this uncertainty is reason for action and not inaction, as we cannot afford the risk of experiencing the worst of the predicted outcomes. Meeting the challenge of providing secure and clean energy resources will be essential to addressing the problems of climate change.

The current engineering options fall into three categories: generating electricity from renewable energy resources; finding ways to make fossil fuel a cleaner source of energy; and reinventing the nuclear dream which promised clean, cheap energy in the mid-twentieth century. What follows is a brief look at the advantages and disadvantages of each of these solutions.

Harnessing solar energy is one way to provide plentiful renewable energy, especially in developing countries. In the mid-2000s US National Academy of Engineering carried out a project 'Grand Challenges in Engineering', setting out to investigate the major goals for engineers in the twentieth century and to identify those seen as the most important. In a web-based poll, the project of making solar energy economical was the runaway winner from a list of challenges that included preventing nuclear terror and engineering medicines. Solar energy is abundant – the earth receives far more energy from the sun than its inhabitants use, so if we could harness just a fraction of that energy, then we could have plentiful renewable carbon-neutral energy.[2]

Current technologies for converting solar energy into electricity do not deliver the amount of power needed at a reasonable cost. However, technologies are in development that may change this, as the area of plastic electronics holds some hope for producing much cheaper solar panels. And in the absence of such technologies solar energy can be used quite efficiently to heat water directly. Other problems with solar energy are the fact that it can only really be generated in abundance at certain locations and, as for any mode of electricity generation, storing it and transporting it across long distances is a problem – electricity is very difficult to store for significant periods of time. But solar energy is a potential prospect for the countries in great need of energy solutions, such as Africa. And Africa could even end up serving the energy needs of Europe with solar energy produced in the Sahara region and distributed by creating a transcontinental 'supergrid'. With investment in solar technology electricity distribution systems, solar power could serve the energy needs of a large part of the world.

The energy of the sun can be harvested in a number of ways – if not directly, then from plants that use sunlight to create their own energy stores through photosynthesis. The development of biofuels (fuels developed from plants rather than fossil fuels derived from oil, coal and gas) can help to address another problem of our energy use. Oil fuels most forms of transport and is making a huge contribution to rises in greenhouse gasses.

Biofuels have been advertised as promising an alternative to oil, with products such as ethanol presenting a cleaner alternative to petroleum. This is not a new technology; the Model T Ford was capable of running on ethanol or gasoline, but gasoline won out as the fuel of choice because it was comparatively cheap.

There are significant costs attached to developing biofuels to power transport, particularly in terms of the land used for the crops that would be converted into biofuels. There is the problem that land in some of the poorest countries is at risk of being diverted from food crops to those that would be used for biofuels, leading to increases in food prices and constraints on supply. The development of biofuels is therefore limited by significant constraints – it may be a more sustainable resource than oil, but it is still not freely or even abundantly available. Biofuels become a better prospect for alternative energy if controls are introduced to ensure that they are derived from plants that do not displace essential food crops. They also offer promise where they are developed from waste – used cooking oil and even the by-products of chocolate manufacturing have been used to power motor vehicles.

Wind energy, like solar, offers the promise of harnessing a free, natural resource to generate electricity. Over recent decades its use has grown significantly, but it still produces only a fraction of most countries' energy supply. One barrier to the expansion of wind power is social – some people object to wind farms as blights on the landscape, and obtaining permission to build them can be a challenge. However, off-shore wind farms that are located in the sea are likely to meet less resistance. But an additional problem is that wind energy could not currently provide the dominant source of energy in most countries because it is not a constant resource; if the weather changes, the electricity supply drops or even disappears. Again, the problem of storing electricity means that the intermittency of wind, and the fact that it can only produce significant amounts of electricity in particular locations, means that the prospects for capitalising on it are constrained. However, as a clean energy that is technologically simple to generate, wind energy holds a great deal of promise.

Coal has long been a favoured means of producing electricity because the process of power generation from burning coal is easy to control, meaning that more or less electricity can be generated according to demand. It provides a reliable means of energy generation through tested technologies, is relatively cheap and, according to current estimates, existing coal reserves should be sufficient for at least a century (though such predictions are highly sensitive to changes in demand). However, coal is a dirty fuel and this means of electricity generation is responsible for a significant proportion of the world's carbon emissions. So what can be done to make it work for us without damaging the environment? Carbon capture and storage, or carbon sequestration, is the process of capturing the carbon dioxide released by burning coal in power stations, and entombing it underground. Thus carbon would be treated similarly to nuclear waste, securely stored where it cannot escape into the atmosphere. A lot of hope was invested in this technology before it was shown to be viable at commercial scales, though early demonstration projects were carried out by the German energy firm RWE and SaskPower in Canada. This is an important technology because, if coal-fired power stations continue to be built across the world, some action has to be taken to reduce their impact on the environment. However, it is not without risk. Even if the technolgy matures, there are concerns about the long term reliability of storage methods and it is not known what would happen if huge amounts of carbon were suddenly released all at once should they fail.

It is just this kind of concern that dogs nuclear power. Nuclear power was once thought to be the future of electrcity generation, providing cheap and plentiful energy. Nuclear energy provides the overwhelming majority of electrical power in France, but in other countries it has proven unpopular and initial flurries of investment in nuclear energy were not repeated. This has been largely due to nuclear accidents or the threat of them, and concern about the as yet unsolved problem of storing nuclear waste. However, interest in this technology has recently been renewed with countries such as the UK and US planning to build a new generation of nuclear plants and others making their first forays into this area. One attraction is the fact that nuclear power promises a secure and controllable supply of electricity; with its own nuclear resource, one country need not be beholden to another to meet its energy needs, as is the case for some fossil fuels. In politically unstable times this is a compelling factor in choices over energy supply. Another major bonus is that it is a low-carbon form of electricity generation and some people argue that urgent investment in this area is needed to curb carbon emissions. Yet it remains a controversial technology, as there is no way of avoiding the creation of radioactive waste that remains hazardous for millennia. The issue to address here – and this is a social and political issue as much as an engineering problem – is whether the long-term risks posed by nuclear waste are greater than the risks posed by climate change. Unless there is a step-change in the amount of electricity produced by renewable energy it is very likely that nuclear power will remain a significant element in the electricity generating portfolio.

Some however see the future of nuclear energy as lying not with existing methods of nuclear fission, but in nuclear fusion. The energy of the sun is created through nuclear fusion – when two particles fuse together and release energy. If we could control this proccess and use the energy released to generate electricity, huge amounts of power could be created from small amounts of fuel without producing harmful waste. So far, reliable methods for harnessing energy in this way have been elusive but research continues into finding a means for generating the kinds of sustained fusion reactions needed to produce electricity. The ITER project, an international research programme based in France, is focused on proving the technical feasibility of producing power through nuclear fusion. This has the potential to be a revolutionary technology, and research in this area is important, but pressing needs mean that more proven energy propositions have been prioritised. International collaboration, as in the ITER project, is essential to allow many countries to have some stake in this technology without committing huge proportions of their energy budget.

Clearly none of these options can provide the sole global solution to our energy challenges, nor even the sole option for any single country. Most countries are likely to exploit a mixture of the technologies, with some countries able to exploit solar energy to a larger extent, others wind, and others carbon capture and storage. Choosing energy supply is a political decision, as it is crucial to a country's stability and security, but it must be informed by engineering reality. Engineers must engage in energy debate, clearly demonstrating the virtues and downfalls of each energy supply option and engage in research and demonstration projects where needed.

Engineering the climate

While significant efforts are being invested in finding a secure means of low-carbon energy generation, there are other strategies for addressing the effects of climate change. One of these is to directly intervene in the earth's climate in order to control the effects that increased carbon in the atmosphere have on global temperatures. Geo-engineering is an emerging area of engineering activity focused on effecting a change in the earth's climate.

There are two ways that engineers can alter the climate and attempt to halt the progress of climate change. One is to capture the carbon that has already been released into the atmosphere to stop its contribution to the greenhouse effect; the other is to reduce the amount of warming caused by sunlight. An example of the first tactic is ocean fertilisation. This involves encouraging the growth of phytoplankton in the oceans, which absorbs carbon dioxide as it flourishes. Some of this plankton will then die off and sink to the sea bed, where the carbon that it has absorbed will rest.

Areas of the earth can also be shaded from the heating effects of sunlight. Following the eruption of certain volcanoes, a cooling of the earth was observed in the vicinity of those eruptions. This was attributed to the release of particulates into the stratosphere. Directly releasing particulates such as sulphates at this altitude could have the same effect.

So what are the prospects for such technologies, and are they the right solution to climate change? Both forms of geo-engineering carry unpredictable risks. Changing the balance of life in the ocean could have unknown and potentially disastrous effects on a fragile ecosystem, and the effects of increasing levels of phytoplankton could be difficult to reverse. Directly intervening in the earth's climate could also have irreversible effects that we cannot foresee. Geo-engineering is a huge gamble. Engineers and politicians may conclude that this should only be a last ditch option. If we can succeed in slowing climate change by reducing our emissions this is a far safer path to take.

Conclusions: the limits of engineering?

In an earlier chapter, I likened the proliferation of engineering disciplines to the evolution of species. Engineering itself has provided the leap that has allowed humans to transcend the process of natural evolution. No longer do we adapt to fit our ecological niche, but we change the world around us to suit our needs, to make up for our weaknesses. Now it is the physical world that evolves to fit our limitations.

We have come to depend on engineering and technology to solve life's inconveniences, but it seems that in doing so we have changed the world in such a way that it may no longer continue to provide a comfortable habitat. Changes made in densely populated parts of the world have changed the earth's most undeveloped landscape – the Polar Regions. They could also have a profound effect on the weather patterns on which we depend for a tolerable climate and habitable land. Halting the process of global warming requires immediate action, and engineering will have a key role in this. We face big problems and we need big solutions of the kind that only engineering can provide.

However, we also need to consider that technological solutions are not the only way forward. The idea that we might resort to something as drastic as geo-engineering to deal with climate change is cause for concern. It would no doubt pose its own risks and would offer no guarantee of sustained effectiveness. Such strategies might be essential in a worst-case scenario, but there is danger in seeing them as a silver bullet for solving our climate problems. As every engineer should know – there is no silver bullet for any problem.

The best solutions to our current problems involve both continuing to adapt the world around us and undertaking to adapt our own lifestyles. Engineers work in and with society – any engineering solution to a problem has a human element. This means that we have to play a part in helping engineers to reduce the carbon emissions that we are responsible for. There needs to be a change in behaviour – less dependence on cars, less use of aviation, and less unchecked use of heating and cooling in our homes. This will make a huge impact on reducing carbon emissions and decreasing the burden on our infrastructure.

The wider public has another role to play in helping engineers to fix the climate problems that we have created. Engineering can do a lot to 'save the world', but there are a number of ways that this can be done, as this chapter has shown. Making these choices is not a matter of selecting on the basis of technological considerations alone, it is a matter of deciding what risks we are willing to take – whether we would rather take the risks posed by novel technologies, or the risks posed by climate change. It is also a matter of deciding whether we are willing to change our behaviour or whether we prefer to maintain our lifestyles and invest significant money, effort and hope into finding engineering solutions to climate change. Engineers are members of the wider public, and they should be actively involved in these decisions. Engineers may have changed the world, but the choices of societies and individuals play an important part in shaping its future.

Notes

Chapter 1

1 Bertrand Gille tells the story this way in *The Renaissance Engineers*.
2 Quoted by Sunny Auyang in *Engineering – an Endless Frontier*.
3 Sunny Auyang, *Engineering – an Endless Frontier*, p. 3.
4 Henry Petroski, *Invention by Design*, p. 2.
5 'Who Is Jonathan Ive?' in *Business Week*, 25 September 2006.

Chapter 2

1 Barbara Lane, of the engineering firm Arup, was awarded The Royal Academy of Engineering Silver Medal in 2008 for her work on structural fire analysis. Her research on the response of buildings like the Twin Towers to fires led her to develop an approach to the design of tall buildings which will make them intrinsically more fire resistant.
2 See chapter 10 of Henry Petroski's *Invention by Design* for the story of how the methods used in the Crystal Palace were vital for the design of tall skyscrapers.
3 Quoted in Victor Smith, *The Sydney Opera House*.
4 The 'i-Snake' robot was developed by a team based at Imperial College London.

Chapter 3

1 FLUENT by Ansys is the industry standard complex fluid dynamics software package for performing simulations.
2 Dym and Little, *Engineering Design: A Project Based Introduction*, p. 6.
3 Adams, *Flying Buttresses, Entropy and O-Rings*, p. 87.

4 James Dyson, *Against the Odds: an Autobiography,* p. 7.
5 Quoted by Sally Dugan, *Men of Iron.*

Chapter 4

1 See Sandy Cairncross: 'Water and waste: engineering solutions that work', *Engineering Change: Towards a sustainable future in the developing world.*
2 For the full story of the Moog synthesiser, see Trevor Pinch and Frank Trocco's *Analog Days: the Invention and Impact of the Moog Synthesiser.*
3 Henry Petroski, *Invention by Design*, p. 187.
4 See *Environmental Values* by John O'Neill, Alan Holland and Andrew Light for a discussion of this kind of issue.
5 Both codes were retrieved from the institutions' websites in August 2008. The IEEE code was approved by the IEEE Board of Directors in February 2006.

Chapter 5

1 I do not wish to suggest that train station staff are sometimes inhuman, but it is increasingly common to find unmanned stations where tickets are bought only from automated machines.
2 See Sandy Cairncross: 'Water and waste: engineering solutions that work', *Engineering Change: Towards a sustainable future in the developing world.*
3 A detailed account of this story was given in the 2008 Lubbock Lecture, presented by Lord Browne of Madingley at Oxford University.
4 For an insight into an engineer's perspective on the nature of consciousness, see Igor Aleksander, *The World in My Mind, My Mind in the World: Key Mechanisms of Consciousness in People, Animals and Machines.*

Chapter 6

1 See M. Masoon Staneskai and Heather Cruikshank, 'Engineering, wealth creation and disaster recovery: the case of Afghanistan', in *Engineering Change: Towards a sustainable future in the developing world*.

2 Carbon neutrality is a popular term, but its use is often a little lazy. While harnessing energy from the sun does not itself produce carbon emissions, the process of manufacturing solar panels is likely to produce some emissions, and the same goes for wind turbines and other renewable energy generating tools.

Further reading

Walter G. Vincenti, 1990: *What Engineers Know and How They Know it – Analytical Studies from Aeronautical History*. Baltimore, Johns Hopkins Press.

Henry Petroski, 2004: *Pushing the Limits – New Adventures in Engineering*. New York, Alfred A. Knopf.

Sunny Y. Auyang, 2004: *Engineering – an Endless Frontier*. Cambridge MA, Harvard University Press.

James L. Adams, 1991: *Flying Buttresses, Entropy and O-Rings: The World of an Engineer*. Cambridge, MA, Harvard University Press.

Bibliography

Aleksander, I. 2007. *The World in My Mind, My Mind in the World: Key Mechanisms of Consciousness in People, Animals and Machines*. Exeter, Imprint Academic

Alvarado, R. V. 2002. *Thomas Edison (Critical Lives series)*. Indianapolis, Alpha

ASME International History and Heritage. 1997. *Landmarks in Mechanical Engineering*. Purdue University Press

Ball, P. 1997. *Made to Measure: New materials for the 21st Century*. Princeton, Princeton University Press

Baker, W. J. 1970. *A History of the Marconi Company*. London, Methuen

Batchelor, R. 1994. *Henry Ford: Mass Production, Modernism and Design*. Manchester, Manchester University Press

Berners-Lee, T. 2003. *Weaving the Web*. London, Texere

Brindle, S. 2005. *Brunel: The Man Who Built the World*. London, Phoenix

Cadbury, D. 1996. *Seven Wonders of the Industrial World*. London, Harper Perennial

Cadbury, D. 2006. *Space Race – The Battle to Rule the Heavens*. London, Harper Perennial

Davis, M. 2004. *Thinking Like an Engineer – Studies in the Ethics of a Profession*. Oxford, Oxford University Press

Dugan, S. 2003. *Men of Iron: Brunel, Stephenson and the inventions that shaped the modern world*. London, Channel 4 Books/MacMillan

Dym, C. L. and Little, P. 2004. *Engineering Design: A Project-Based Introduction*. Chichester, Wiley

Dyson, J. 1997. *Against the Odds: an Autobiography*. London, Orion Business Books

Emmerson, G. S. 1973. *Engineering Education: A Social History*. Newton Abbott, David and Charles

Fagerberg, J., Mowery, D. C., and Nelson, R. R. (eds). 2005. *The Oxford Handbook of Innovation*. Oxford, Oxford University Press

Ferguson, E. S. 1992. *Engineering and the Mind's Eye*. Cambridge, MIT Press

Gille, B. 1996. *The Renaissance Engineers*. London, Lund Humphries

Gordon, J. E. 1978. *Structures: Or Why Things Don't Fall Down*. London, Penguin

Greaves, W. F. and Carpenter, J. H. 1978. *A Short History of Mechanical Engineering*, 2nd Edition. London, Longman

Guthrie, P., Juma, C. and Sillem, H. 2008: *Engineering Change: Towards a sustainable future in the developing world*. London, The Royal Academy of Engineering

Harvie, D. I. 2004. *Eiffel: The Genius Who Reinvented Himself*. Stroud, Sutton Publishing

Hughes, T. 2004. *Human Made World*. Chicago, University of Chicago Press

Humphreys, K. K. 1999. *What Every Engineer Should Know About Ethics*. New York, Marcel Dekker

Ihde, D. 1993. *Philosophy of Technology, an Introduction*. New York, Paragon House

Israel, P. 1998. *Edison: A life of invention*. New York, Wiley

Jones, P. 2006. *Ove Arup: Master Builder of the Twentieth Century*. Yale,Yale University Press

Luegenbiehl, H. C. 2004. 'Ethical Autonomy and Engineering in a Cross-Cultural Context', *Techné*, vol. 8

Matthews, C. 1988. *Case Studies in Engineering Design*. London, Arnold

Mitcham, C. 1994. *Thinking Through Technology – The Path Between Engineering and Philosophy*. Chicago, The University of Chicago Press

Morris, P. R. 1990. *A History of the World Semiconductor Industry*. London, Peter Peregrinus

Nakazawa, H. 1994. *Principles of Precision Engineering*. Oxford, Oxford University Press

Parsons, W. B. 1939/1967. *Engineers and Engineering in the Renaissance*. Cambridge, MIT Press

Petroski, H. 1992. *To Engineer is Human: The Role of Failure in Successful Design*. New York, Vintage

Petroski, H. 1996. *Invention by Design – How Engineers get from Thought to Thing*. Cambridge, Harvard University Press

Pinch, T. and Trocco, F. 2002. *Analog Days: the Invention and Impact of the Moog Synthesiser*. Cambridge, Harvard University Press

Pitt, J. C. 2000. *Thinking about Technology – Foundations of the Philosophy of Technology*. New York, Severn Bridges Press

Pitt, J. C. 2001. 'What Engineers Know'. *Techné* 5:3, 17–30

Pool, R. 1997. *Beyond Engineering: How Society Shapes Technology*. Oxford, Oxford University Press

Rogers, G. F. C. 1983. *The Nature of Engineering: a Philosophy of Technology*. London, Macmillan

Sample, I. 2007. 'Profile: Jonathan Ive', *Guardian*, Friday January 5

Smith, V. 1974. *The Sydney Opera House*. Sydney, Paul Hamlyn

Vise, D. A. 2005. *The Google Story*. London, Pan Books

Watson, G. 1998. *The Civils: The Story of the Institution of Civil Engineers*. London, Thomas Telford

Whitbeck, C. 1998. *Engineering Ethics in Practice and Research*. Cambridge, Cambridge University Press

'Who Is Jonathan Ive? An in-depth look at the man behind Apple's design magic', *Business Week*, September 25 2006

Common Methodologies for Assessing Risk. 2003. The Royal Academy of Engineering

Creating Systems that Work. 2007. The Royal Academy of Engineering

The Economics and Morality of Safety. 2006. The Royal Academy of Engineering

Risks Posed by Humans in the Control Loop. 2003. The Royal Academy of Engineering

The Societal Aspects of Risk. 2003. The Royal Academy of Engineering

Index